WRITING FOR SCIENCE

Writing

FOR SCIENCE

Robert Goldbort

Yale University Press

New Haven & London

Designed by Nancy Ovedovitz and set in Times Roman type
by The Composing Room of Michigan, Inc. Printed in the
United States of America by Vail-Ballou Press,
Binghamton, New York.

Library of Congress Cataloging-in-Publication Data
Writing for science / Robert Goldbort, 1949–
p.   cm.
Includes bibliographical references and index.
ISBN-13: 978-0-300-11551-2 (cloth : alk. paper)
ISBN-10: 0-300-11551-2 (cloth : alk. paper)
ISBN-13: 978-0-300-11793-6 (pbk. : alk. paper)
ISBN-10: 0-300-11793-0 (pbk. : alk. paper)
1. Technical writing.   2. Communication in science.
I. Title.
T11.G626   2006
808′.0665—dc22   2006012742

A catalogue record for this book is available
from the British Library.

The paper in this book meets the guidelines for permanence
and durability of the Committee on Production Guidelines
for Book Longevity of the Council on Library Resources.

10 9 8 7 6 5 4 3 2 1

To Joanne
and to our treasures
Raechel, Jonathan, Julia, Sarah

CONTENTS

6 SCIENTIFIC VISUALS   174

7 SCIENTIFIC PRESENTATIONS   194

8 SCIENTIFIC DISSERTATIONS   213

9 SCIENTIFIC JOURNAL ARTICLES   240

10 SCIENTIFIC GRANT PROPOSALS   271

Contents

PREFACE

Although *doing* science is at the heart of discovery, the effort would have very limited consequence in the long term without *writing* science. As a social enterprise that depends on collaboration, scientific inquiry requires its practitioners to write on a regular basis. From time to time, some members of the scientific community have been critical of the overall quality of writing by researchers. If scientists do indeed write less effectively than writers in other professions, at the root of that circumstance may be the sentiment that time spent writing is far less important than time spent doing research. Shouldn't fussing over writing and language be left to the "literary" writers? Won't the results, after all, speak for themselves? This book stands with other scientific writing guides in responding "no" to such questions. The profound impact of science in our world demands special care and the most rigorous standards for communicating its outcomes to the multiple constituencies affected by their implications and applications.

This book is intended to help students and scientists maximize the effectiveness of the writing that they must do during their education and professional life. One archetypal image of a scientist at work is that of an engrossed observer who carefully records experimental findings in a laboratory notebook. Although meticulous notekeeping is at the core of scientific research, such an image does not fully represent the role in science of writing. Collectively, the chapters in this book demonstrate how scientists' writing ranges widely in form and purpose. Moreover, whatever the form, purpose, or audience of scientific writing, the one common denominator is the demand that researchers use language in the highly formalized manner that accords with empirical senses of knowledge, truth, and precision.

With a number of fine scientific writing guides already available, what makes this book different? First, no other current guide is as comprehensive. There are guides devoted solely to grammar and usage, papers, theses, and proposals, but no single reference covers the full gamut—including lab notes, workplace communication, undergraduate reports, and scientific documentation—with chapter-length treatments of each. Second, the in-depth approach in chapters, versus the short-entry style of handbooks or manuals, allows the use of extended examples. An illustrative thread that runs through the book is the area of alcohol studies, and the chapters on dissertations, articles, and proposals use specific documents extensively for continuity and depth of coverage. Third, the book is unique in its number and rigor of examples, centering on the various forms, purposes, and features of scientific writing. The comprehensive treatment of the various kinds and purposes of scientific writing, together with the quantity, rigor, and highlighting of examples, make this book an important complement to the current array of writing guides for students and working researchers.

A book of this nature must of course rely to a considerable extent on those authors whose work has provided a foundation in the discipline that allows others, like myself, in turn to make their own contributions. Their publications, many of which are cited throughout the text, have been guiding lamps for my own thinking and teaching. David Locke, for instance, offers the important notion of "science as writing" in his book by that title; Michael Katz's *Elements of the Scientific Paper* demonstrates the effective use of extended examples; and F. Peter Woodford's *How to Teach Scientific Communication* emphasizes the importance of teaching scientific writing to graduate students. These and numerous other texts on scientific writing provide invaluable historical, theoretical, and practical insights in a discipline that is still relatively young and growing in its scholarship.

I could not close these prefatory remarks without acknowledging those individuals who contributed in one incalculable way or another to my being able to complete this book project. I am indebted to three mentors along my academic journey in biology and English—Professors Sheena Gillespie, Carl Schneider, and Stephen Tchudi—whose passion for their work and personal encouragement demonstrated to me how teaching, learning, and writing are so inextricably connected to who we are as human beings. I am particularly grateful to Professor Tchudi, my dissertation adviser, for his invaluable insight that one's syllabus is also one's book. I also wish to extend my deep gratitude to two faculty members in the Department of Life Sciences at Indiana State University, Michael Angilletta and Steven Lima, who offered their expertise and writing samples for Chapters 8, 9, and 10. Those chapters would have been far less useful and interesting without their generosity in sharing their outstanding work. For assistance with visuals, I thank Sarah Edwardson of the Center for Instruction, Research, and Technology at Indiana State University. At Yale University Press, I am fortunate to have had the constant support of Jean Thomson Black—who believed in this project from the moment she read the early chapters—and grateful to Phillip King for his keen editing. Finally, no words will suffice to express gratitude to my family. My parents, Jaime and Victoria, encouraged my exploration of the cultures of science and English. Without my wife, Joanne, my everything—both throughout our decades of intertwined growth and during my long hours of isolation beginning in the spring of 2003—this book simply would not be a reality.

SCIENTIFIC ENGLISH

"When *I* use a word," Humpty Dumpty said, in a rather scornful tone, "it means just what I choose it to mean—neither more nor less."

"The question is," said Alice, "whether you *can* make words mean so many different things."

"The question is," said Humpty Dumpty, "which is to be master—that's all."

—Lewis Carroll, *Through the Looking Glass and What Alice Found There*

LANGUAGE AS A TOOL OF SCIENCE

Scientific English is a number of things. It is a communication tool, a culture of writing, and a plain and readable manner of writing with specific compositional strategies and uses of language—all of which permit the community of scientific researchers to conduct its professional affairs. In desiring essentially to be masters of their own language, scientists rely on narrowly restricted uses of words. The linguist Leonard Bloomfield has explained the benefits of this scientific way of communicating: "The use of language in science is specialized and peculiar. In a brief speech the scientist manages to say things which in ordinary language would require a vast amount of talk. His hearers respond with great accuracy and uniformity. The range and exactitude of scientific prediction exceed any cleverness of everyday life: the scientist's use of language is strangely effective and powerful. Along with systematic observation, it is this peculiar use of language which distinguishes science from non-scientific behavior."[1]

The primary purpose of this chapter is to delineate and illustrate the unique linguistic values that the scientific community places on the way it uses words for conducting its activities and achieving its goals. How do scientists use language? How does using English (or for that matter any other language) scientifically differ from other uses to which language may be put? The explanations in the first sections of this chapter on the professional, historical, and philosophical contexts that define scientific uses of language will be followed in the remaining sections by actual examples of scientific English in practice. Defining scientific English risks making hard and fast distinctions about the way language works or among the things that humans do with it. Therefore, making general pronouncements in an attempt to draw lines between kinds of uses of language is bound to be met, on one intellectual front or another, by resistance. Language study today is a complex field that utilizes multiple perspectives, including those of composition and rhetorical theory, communication, cognitive psychology, sociology, anthropology, and neurobiology.

All that said, there are nonetheless practical distinctions to be drawn. In practice, it is safe to say that a basic criterion for defining scientific uses of language is that of the user's intent. Scientists use language strictly and narrowly as a communication tool. This distinguishing intention of communication shapes the professional culture and compositional style of scientists as writers. The communication model of using language suggests that words are merely physical objects or mechanical tools. Applied to scientific language, this rather simplistic view limits the role of words to something like conveyor belts in automated factories, delivering to their readers units of objective information derived from and in the service of the equally objective methods of scientific inquiry. In contrast, non-scientific uses of language like those in the literary world give prominence to personal and subjective expression. In actuality, the use of scientific language has inherent biases and subjectivities that, however desirable it may be to eliminate them, are an inescapable dimension of the human presence in written texts. Here we have, then, the key distinguishing criterion: the priority that scientists as writers, as users of the English language, give to the objective information that words impart. This central priority of communicating information demands that scientists use the tool of language responsibly and effectively to serve a scientific purpose, with the aim of convincing their intended readers of that purpose's value. There is a wide range of documents that scientists can use for achieving this effect with their discourse.

The fundamental point to keep in mind is this: any attempt to understand scientists as writers must begin with the observation that their work and their documents depend vitally upon language. From note taking to publishing and teaching, language is the tool that gives sense to scientific activity. Whatever scientists do and observe, everything they come to know or to hypothesize, is mediated through language: "There is no real world that scientists know independently of the linguistic, graphic, and mathematical formulations by which they conceive it," one author on scientific writing has underscored.[2] Without the resources of language, the scientific enterprise would not progress for long. The mathematician Jacob Bronowski asserted that "the method of science, the objectification of entities, abstract concepts, or artificial concepts like atoms, is in fact a direct continuation of the human process of language, and that it is right to think of science as being simply a highly formalized language."[3] What does "a highly formalized language" mean? What are the specific and practical rules of scientific English? To understand what it means to use scientific English effectively—at the level of words, sentences, and paragraphs—it is helpful to understand what scientific English is in its broader contexts: What are its scope, aims, and linguistic qualities? What are the professional relationships among scientist-writers, their documents, and their intended readers? What is the historical origin of the scientific attitude toward language? It is only through the lens of the historical evolution of modern science's view of language that the effectiveness of today's scientist-writers can be gauged. Therefore, the specific practical examples given later in the chapter will make more sense in light of this modern linguistic evolution. The basic nature of scientific English can be illuminated within two basic contexts: first, as constituting a practical communication framework, a culture of writing, founded on certain professional aims and purposes, and second, as a utilitarian attitude that cultivates an ethic of plainness in the use of language for scientific ends.

THE COMMUNICATION RANGE OF SCIENTIFIC ENGLISH

The sense of scientific English as a tool for organized communication is not disconnected from the classical Greek and Roman philosophies of discourse that two millennia later have come to shape the way college English, especially report writing, is taught today. Expository writing in any discipline has roots in Aristotle's methods for supporting a thesis or in Cicero's way of dividing an oration that easily translates into the various parts of a research report. Therefore, much of traditional college English is also part of what defines scientific English. Also apparent, however, is that a relative newcomer to the academic world—Francis Bacon's experimental science—brought along new and scientifically plain ways of using language for new purposes in new documents for new readers. Scientific English, then, has its own professional culture of writing. Its historical evolution since Bacon actually has extended rather than rejected Aristotle's and Cicero's contributions to the effective use of language. The Baconian outlook became an irrepressible impetus toward the emergence of the ethic of mathematically plain scientific communication.

Given the prime motive of communication in the culture of scientific writing, several questions naturally follow: To communicate what? Why? To whom? In what forms and styles? The geneticist Bentley Glass observed that there are "at least five distinct obligations" shared by scientists in their professional communication:

- publishing their methods and findings truthfully, clearly, and fully so that they can be verified and extended by fellow researchers;
- disseminating their findings more widely through abstracting and indexing media;
- writing critical reviews that synthesize current knowledge in their field;
- sharing their knowledge and its practical implications with the public;
- teaching what they know to future generations of scientists.[4]

To Glass's list, one may add the writing of laboratory notes on research methods and outcomes, proposals of research to acquire grant funding, and daily on-the-job communication. Given all these goals, we can identify six basic kinds of purposes that researchers have when they write particular documents for particular readers in order to achieve those purposes effectively: recording and archiving, professional exposition or dissemination of research results, teaching, job duties, seeking financial resources, and informing citizens (Table 1.1). In scientific activity itself, the most immediately important uses of language occur in making a reliable and permanent record or archive of research methods, outcomes, and conclusions (see Chapter 2). The next professional purpose for researchers is to share their work with peers through publication. Beyond these prime archival responsibilities—which allow the profession to advance in the collaborative spirit it requires—scientists also must share their knowledge in various forms with a range of reader-

Table 1.1 Purposes, types, audiences, and styles of scientific writing

Purpose	Document Types	Intended Readers	Linguistic Style
Recording and archiving	Laboratory notes, with other preservable forms of documentation, such as equipment, printouts, photos, and special artifacts for verifiability	Self; research collaborators; workplace supervisors	Informal to highly formal notations in arcane shorthand; lab jargon
Professional exposition and synthesis	Scholarly articles and books; abstracts; notes and visual media for conference papers and seminars; letters; e-mail	Researchers in same or related field	Highly formal, with heavy use of jargon
Teaching	Textbooks, syllabi, electronic slides, Web-based infor-mation, and other pedagogical materials	Students at all levels	Moderately to highly formal, with parallel range in jargon
Performing job duties	On-the-job communi-cations, including e-mail, letters, mem-oranda, meeting minutes, and activity or progress reports; internal and external	Research associates, colleagues, and administrators	Informal to highly formal; low to high level of jargon
Seeking research resources	Grant proposals to government agencies, corporations, and philanthropic foun-dations	Granting agency officials; peer reviewers	Highly formal; moderate to heavy use of jargon

(continued)

Table 1.1 (*continued*)

Purpose	Document Types	Intended Readers	Linguistic Style
Informing citizens	Articles, essays, and books; special letters; Web-based material; creative forms; expert testimony and other consulting documents	General public; special-interest groups	Formality and jargon low to moderate

constituents. These interested readers range from students and fellow researchers to public officials and citizens. Each of the important purposes in scientific writing calls for a particular nuance in the basic manner of using scientific English, in how formal or detailed the communication may need to be. A culture of writing also means a culture of readers. The particular choices that scientists make as writers must be guided by assumptions about their readers.

It is not enough, then, for effective and responsible scientist-writers to know their subject. They also must know a document's readers; for example, how much do *they* know about the subject? Is the document for a research supervisor, a journal, a public official? How should a document's technical formality and style be adjusted for its reader(s)? Do the writer's intentions match the reader's expectations? Consider any given document mentioned in Table 1.1 in light of this question: What would the reader expect? Scientists do write for their all-important and diverse readers with their range of expectations. The professional standards for *doing* science are reflected in the strict standards and practices for *writing* science. The modern scientific community's culture of writing also demands a unique sense of plain language. This sense of scientifically plain English is both a cause and an effect of the rise of the experimental sciences inspired by Francis Bacon's revolutionary new senses of human "knowledge," of "reality," and of "truth." One prefatory caveat: Although the historical evolution of the notion of modern scientific language as thoroughly objectified is well documented, today scientific language is more accurately seen as also having subjective elements—psychosocial and political—that may affect its ultimate truth value. Before considering that humanized dimension of scientific language, however, a broader sense of its history is necessary to explain its Baconian roots.

THE LEGACY OF SCIENTIFICALLY PLAIN ENGLISH

The truly monumental achievement of the so-called father of modern science, Francis Bacon, is twofold: First, he set human learning on a new course that resulted in what today we call modern science, which seeks to advance human understanding through observing and manipulating our natural and physical world. Sometimes we refer to this modern method of study as the "experimental" or "hard" or "exact" sciences—like biology, chemistry, and physics—with the primary sense of the word "research" as inquiry that goes on in a laboratory setting. Second, and just as important, Bacon set the new communication standard or ethic of linguistic plainness that empowered his new scientific program to achieve the grand success it has enjoyed to this day. In short, Bacon at once provided both the method and the language of modern science. What, then, is the linguistic revolution that brought us scientifically plain English? What does it mean to be scientifically plain? What are the specific qualities of plain writing that are expected in scientists' writing?

OLD AND NEW USES OF LANGUAGE: WORDS VERSUS THINGS

In Bacon's view, traditional or past uses of language—stilted, convoluted, clouded with subjective and flowery language—were no longer adequate for advancing human understanding. At the dawn of the seventeenth century, as he laid out a new and bold scientific enterprise, Bacon also chastised those who "hunt more after words . . . than after the weight of matter, worth of subject, soundness of argument, life of invention, or depth of judgment." With the rise of modern science, the dominance of the old attitude of taking pleasure in linguistic artistry and subjective thoughts for their own sake—as in literary writing—was displaced by the Baconian ethic of linguistic utility: how effectively the words serve their readers in delivering "real" knowledge with clarity and exactness. Whereas the traditional linguistic style reveled in subjective ambiguity, the new one was to be utterly and objectively plain in the service of true learning. When Bacon's dream of a modern research institution became a reality in the Royal Society of London, the society's members officially resolved "to reject all the amplifications, digressions, and swellings of style: to return back to the primitive purity, and shortness, when men deliver'd so many *things,* almost in an equal number of *words.* They have exacted from all their members, a close, naked, natural way of speaking; positive expressions; clear senses; a native easiness: bringing all things as near the Mathematical plainness, as they can." Rather than a return to some golden era of "natural" writing, however, Bacon's was a new and future-oriented standard that reflected modern science's forward-looking way of thinking and learning. The essence of the pivotal linguistic revolution that accompanied the modern scientific revolution is the emergence of this new ethic of "mathematical plainness" that values things over words. This Baconian attitude toward language can be translated, or paraphrased, into the following current mantra of scientific plainness: "There should be little figurative language . . . an economy of words . . . intelligible, clear, and unequivocal meanings . . . common words which are closer to material realities . . . no emphasis upon or interest in the mode of expression for its own sake . . . Rhetorical ornaments and sheer delight in language represent a pernicious misplacing of emphasis, and in the end destroy the solid and fruitful elements of knowledge."[5] For scientists, writing that is worth reading has real *things* to offer in mathematically plain language. The utility of scientifically plain English lies in those two fundamental and interconnected features: first, that it has practical material to offer, and second, that it communicates that material plainly so it can be used by the reader.

The key shift in the rise of the new sciences with their new senses of knowledge and truth was in what was meant by "things." Baconian things were not the same as, say, the relatively subjective Aristotelian or Ciceronian things. According to Robert Adolph: "Bacon means by 'things' objective physical reality and its causes, existing before and after the writer's perception of them and independent of him. The Baconian writer, like his ideal researcher, submits his mind to these things, rather than constructing a mental edifice of his own according to some ideal pattern or looking within himself to relate the physical world to his own private concerns."[6] Scientists as writers must offer objective knowledge to their readers in plain language. Scientifically plain writing is objective, simple, precise, concrete, direct, and unadorned, with straightforward constructions and the minimum number of words needed to deliver the document's material things to its readers. Of these pivotal changes in human history, it is rightly put that "no clearer proclamation could be desired of the victory of the new world-picture, the fact world, over the older worlds of traditional feeling. 'Truth' was the exclusive possession of the Real Philosophy."[7] The new language of science focused not on psychological but rather on material reality. The Baconian attitude toward language largely defines the present culture of writing in the community of scientific researchers, wherein words are used in very specific, constrained, highly formalized, and generally impersonal ways that accord with scientific objectivity. The old emphasis on the writer and on artistic language has given way in the past four centuries to the modern scientific emphasis on words merely as neutral conveyors of information for the practical benefit of the reader.

THE PLAIN ENGLISH MOVEMENT AND READABLE SCIENTIFIC WRITING

Since the 1970s and 1980s, and not just coincidentally with the emergence of the computer age and then the information age, the ethic of mathematical plainness in scientific discourse has been at the center of the so-called Plain English Movement. Computers have made it easier both to create and to retrieve vast seas of technical information, which users expect to be reader-friendly. One document designer's definition is not much different from that of the Baconians: "Plain English means writing that is straightforward, that reads as if it were spoken. It means writing that is unadorned with archaic, multisyllabic words and majestic turns of phrase that even educated readers cannot understand. Plain English is clear, direct, and simple." The historical circumstances in the last quarter of the twentieth century sparked a reinvigorated demand for readable technical language. Technical businesses like International Business Machines and General Motors developed plain-writing guidelines for their employees and have supported them with in-house desktop publishing resources. In government, President Jimmy Carter led the way with his signing of Executive Order 12044 on March 24, 1978, part of which required that federal regulations be "written in plain English and [be] understandable to those who must comply with [them]." In the world of public affairs, plain and reader-friendly English is not just more effective for getting the job done; it is also more economically efficient. This reinvigorated call for plainness by the public was accompanied by a widespread interest in theories of document readability.[8]

Defining Scientific Readability

In academic writing, the *Publication Manual of the American Psychological Association* (APA) tells us how to be scientifically clear and "agreeable" for the reader; as to how scientific prose should read, there are plenty of current variations on the Baconian theme of plain and measured English. One experienced editor of scientific books and journals writes: "The beauty of science is in the science, not in the language used to describe it. The beauty of English is its ability, when properly used, to express the most complicated concepts in relatively clear words and to point up the beauty of the science. Successful communication in science involves that magic word, *clarity,* a kissing cousin of *simplicity*." Again, the call in science is for reader-centered writing. In our age of information technology, reader-friendly communication

9

will only continue to grow in demand. The basic principle remains simple: No matter how much information a document may contain, if comprehension of it is blocked by inaccessible or imprecise language then the writing is not much more useful than the pre-Baconian varieties of linguistic ambiguity and opaqueness. Fundamentally, the concept of readability simply places readers at the center of communication, facilitating their decoding of information without making them expend undue time and effort re-reading. Writing readable scientific prose means putting into practice, using various compositional strategies, the principles of objective wording valued by research scientists. The more generalized call of the Plain English Movement for reader-centered writing, with its readability theories, also produced mathematical formulas for measuring how readable a document is.[9]

Measuring Scientific Readability

Readability formulas are designed to measure qualities of writing that comport with a scientific style, with simple, direct, and concise wording. The word-processing software you use probably has a feature to calculate the readability of your writing. Stand-alone style and grammar checkers also have been marketed under such names as RightWriter, CorrectGrammar, Editor, and Grammatik. These programs use readability formulas, such as Flesch-Kincaid, Dale-Chall, Spache, and Gunning, to measure the number of technical words, number of syllables, and length of sentences and paragraphs in a written work. To get a sense of how readability formulas work, try computing the so-called Gunning Fog Index by taking a short technical report and following three simple steps:

1. *Average sentence length (ASL):* Count the sentences in several 100-word samples and divide the total word count by the sentence count.
2. *Percentage of hard words (PHW):* Count the words in your samples that have at least three syllables, excluding proper names, simple compound words (e.g., afternoon, humankind), and verbs with three syllables due to *-ed, -es,* or *-ing* endings (e.g., enriches, extruded).
3. *Gunning Fog Index (GFI):* Add your ASL and PHW from the first two steps and multiply that sum by 0.4. For example, an ASL of 15 and a PHW of 21 adds up to 36, which, when multiplied by 0.4, yields a GFI of 14.4.[10]

The GFI value represents the document's level of difficulty as a grade level, which in this case means that readers should have a grade 14, or college sophomore, reading ability. The various formulas work their magic in different ways. The Dale-Chall formula checks a document's ASL and the number of words that are not on its list of a few thousand words. The Cloze Procedure deletes every fifth word and determines the readability score according to how difficult it is for a reader to fill in the blanks. However much faith one may place in such devices, it should not be surprising that a scientific attitude toward language would lead to experimentation with mathematically objective methods for measuring the readability of formal writing.

Such mathematical devices may be interesting as a benchmark of sorts but have limited practical use where linguistic options must be weighed using human judgment. Readability calculations are fraught with limitations because they are based on simplistic views of how humans write and read and think. Shorter words and sentences, for example, are not always easier to read. An ordinary four-syllable word like "separation" is easier for any reader to understand than a three-letter scientific word like "ohm." Moreover, readability formulas easily can be manipulated to yield higher or lower levels of difficulty by making just a few simple textual revisions. Breaking up a document into a greater number of shorter and simpler sentences, for instance, will reduce the reading grade score without necessarily changing the actual difficulty of the content. Besides content difficulty, readability formulas do not take into account such factors as concrete versus abstract language, specialized technical vocabulary, sentence structure and syntactic complexity, the reader's prior knowledge, clarity of the writer's purpose, logical organization and coherence, integration of verbal and visual information, and document layout and design. The act of reading is a complex human process that involves cognitive, linguistic, cultural, and rhetorical dimensions. Therefore, relying heavily on formulaic devices to measure arbitrarily certain features of a written text is at best somewhat of a simplification. Inevitably, however, any full consideration of what scientific English is must return to the basic truth that using language is indeed a human act that, within traditional practices and purposes, displays individuality and originality of style. Rather than undermining or contradicting the historical observation of scientific language as an objectified entity or tool, highlighting its subjective side simply completes the picture as it has come to be recognized in our time.

THE HUMAN DIMENSION OF SCIENTIFIC ENGLISH

The aim here so far has been to explain the historical view of scientific discourse as a tool that is as facilitative and yet as neutral as a piece of laboratory equipment like a Bunsen burner or a cyclotron. Scientific English is expected to transfer information without interfering with clarity, readability, and utility. As writers, scientists are narrowly restricted to their professional universe of discourse and its lexicon. Computers can perform comforting if crude measures of the mathematical plainness of discourse. The apparatus of scientific English is commonly held to be objective and impersonal to the point where its user is perceived as irrelevant and invisible. The data, such a view holds, speak for themselves. Passive constructions that avoid personal pronouns— "experiments were conducted," for example, versus "I conducted experiments"—are seen as a way for researchers to maintain a heightened sense of objectivity in their writing. It is also nonetheless true within acceptable professional bounds, or even in challenging those bounds, that scientific discourse still reflects a writer's individuality. This individuality is evident in at least three basic ways: First, the personal research style of every scientist is reflected in an individual prose style; second, a researcher may use innovative language or new terminology (neologisms); and third, researchers may make choices that are anchored or tinctured sociopolitically.

THE PERSON IN SCIENTIFIC DISCOURSE

Objectivity in scientific writing does not mean that the writer must sound like a lifeless automaton. The presence in scientific discourse of the writer's persona, while perhaps helpful only for evincing authorial integrity, is unavoidable. The individual character of scientists' writing reflects their human diversity as a professional community. As the physiologist Peter B. Medawar observed: "Scientists are people of very dissimilar temperaments doing different things in different ways. Among scientists are collectors, classifiers, and compulsive tidiers-up; many are detectives by temperament and many are explorers; some are artists and others artisans. There are poet-scientists and philosopher-scientists and even a few mystics. What sort of mind or temperament can all of these people be supposed to have in common? *Obligative* scientists must be very rare, and most people who are in fact scientists could easily have been something else instead." Just as no two scientists, even in the same specialty, conduct their research in precisely the same manner or style, no two scientists sound or "read" the same in their professional writing. The microbiologist Salvador Luria used a musical analogy to comment on the professional significance of a scientist's personal style: "Closely related to the role of imagination in scientific research is the question of style. No two scientists, especially effective scientists, function identically, just as no two violinists play Bach's

Chaconne in exactly the same way. I choose this example advisedly, since both violinist and scientist have limited freedom, the former bound to the score, the latter to a factual context, but within the range of their freedom each performs with a unique personal style. Just as an experienced listener can tell which virtuoso is playing, so an experienced scientist can often tell which virtuoso is the author of an important scientific paper." Luria observed a range in the personal styles of his colleagues' papers from "terse" and "almost whimsical" to "slightly baroque" and "aggressive," noting that each of these scientists "is distinct in style because each is a unique self and projects that self into every aspect of his work." This personal dimension or range of freedom exists without violating the professional ground rules of the scientific community's shared traditions and conventions for communicating science effectively and clearly.[11]

ORIGINALITY AND INNOVATION IN SCIENTIFIC DISCOURSE

It is not only impossible to completely depersonalize scientific writing, but it would also not be desirable. The individuality of scientific discourse is often expressed beneficially in such ways as the use of clarifying (rather than obfuscating) figures of speech like metaphors or analogies and in the creation of entirely new words. Such individual originality in language is a quality that scientific inquiry can ill afford to lack. To the contrary, as one analysis of the subject points out: "No synthesis could ever be achieved, no models postulated, no paradigms established, if science relied wholly upon 'careful observation' for its theories. Model-building requires an inductive leap; carefully recorded examples must be synthesized into a logical premise, and then be further verified and expanded by traditional scientific method. For this, science must exploit the power of metaphor; it must shape its expectations, choose its experiments, and interpret its data in a realm of thought outside the literal world." This does not mean, of course, that anything goes. New language or expressions must stand the test of peer scrutiny and be seen as making scientific sense. These may be relatively simple images like describing red blood cells as "sickled" or, when they agglutinate, as appearing like "a roll of coins," or visualizing a triangular laboratory apparatus as "pie-shaped." Images may be more complex, like that of an atom as a solar system with a nucleus (sun) and with particulate matter like "charms" and "quarks." Use of such language is especially helpful in rapidly developing areas of science. The nineteenth-century physicist James Clerk Maxwell believed that metaphors are not only "legitimate products of science, but capable of generating science in turn."[12]

Innovative language has been helpful in both advancing and communicating scientific knowledge to various audiences, including peer researchers, students, and the public. Consider for instance the potent synthesizing value of seeing the molecular structure of benzene as a hexagonal ring formed by atoms arranging themselves like six snakes connected head to tail, as did German chemist Friedrich Kekulé in the 1860s, solving a fundamental mystery in organic chemistry. Or, a century after Kekulé, college textbooks included the popular biochemical "lock and key" analogy for visualizing the mediating roles of enzymes through a process of coupling with their chemical substrates. In genetic chemistry after the work of Nobelists James Watson and Francis Crick, we all came to know the double helix metaphor for DNA. As the molecular genetics revolution unfolded, that apt metaphor gave rise in scholarship and textbooks to a constellation of terms for describing and explaining DNA's role in a "messaging" model requiring "coding," "transmitting," "transcribing," and "translating" information in the process of gene "expression." As this novel genetic model took hold in the public's imagination, the "genetic code" became an "alphabet" with which to read and reveal the encrypted meaning contained in the many volumes of information contained in our cells' genetic encyclopedia. In evolutionary genetics, Stephen Jay Gould and Richard Lewontin used an architectural metaphor in a 1979 professional paper when they compared the elaborations of natural anatomy to elaborately decorated spandrels—tapering triangular spaces formed when four columns support a dome, as in St. Mark's cathedral in Venice. Gould and Lewontin's purpose was to argue that adaptationists too often look at "secondary epiphenomena," like the decorations on the spandrels, as a cause of natural forms (such as divaricate patterns in mollusks) rather than as a direct effect of structural systems in nature. Such innovative expressions are important linguistic tools that can guide and organize scientific thought and work, and that sometimes make their way usefully into formal scientific exposition. As Luria noted, clarity in scientific prose is not monochromatic. Original and innovative scientific English can indeed be helpful, whether in visualizing a natural structure, explaining a complex phenomenon, or offering a new theory. Creativity in language is not the domain solely of literary art, but rather a common thread across all forms of knowledge making and professional communication.[13]

THE SOCIOPOLITICAL CONTEXT OF SCIENTIFIC DISCOURSE

As much as the authority of scientific discourse depends on a detached objectivity, a factual foundation, and powers of logical reasoning, researchers do not communicate their professional knowledge in a social vacuum. This basic observation is evident in various ways. One major influence on scientific activity and its communication is what Thomas Kuhn described as "paradigms," basic sets of assumptions or perceptions that shape how researchers may design, interpret, or convey their experimental work.[14] The power of such paradigms may shape scientific thought and activity in either positive or negative ways. Paradigms or models of atomic, genetic, and cellular structure and function have permitted the scientific community to work collaboratively toward achieving enormous advances in our understanding. One need only consider in these contexts such areas as laser technology, genomic engineering, DNA forensics, or micro-targeted drug therapies.

Scientific progress may also be thwarted, however, by cultural thinking of the day. One prominent example is the sexually prejudiced science that led widely respected nineteenth-century craniometrists like Paul Broca and Gustave Le Bon to set forth theories of the biological inferiority of women. Their extensive measurements of the size and weight of male and female brains were used to support a priori conclusions, as Gould has shown from his review of the available data. Here is a sample quote by Broca that Gould pointed out: "We might ask if the small size of the female brain depends exclusively upon the small size of her body. Tiedemann has proposed this explanation. But we must not forget that women are, on the average, a little less intelligent than men, a difference which we should not exaggerate but which is, nonetheless, real. We are therefore permitted to suppose that the relatively small size of the female brain depends in part upon her physical inferiority and in part upon her intellectual inferiority." Such theories of biological determinism have heaped similar disparagement on blacks and poor people. Gould concludes that, in fact, the corrected and true differences between the weight of male and female brains likely is negligible "and may well favor women" over men.[15]

A full century after Broca and his disciples arrived at such unjustified conclusions, the neurophysiologist Ruth Bleier cautioned against similarly biased theories. Writing in the 1970s, Bleier asserted that there is a set of questions we may legitimately pose to most fields of research, however objective those fields may seem. "To what degree," she suggested we ask of any research results, "do one's philosophical, political and social biases affect one's scholarship, the questions one thinks to ask of the experimental model, the language one chooses to pose the questions, the nature of the controls one considers relevant, and finally, the openness or breadth of one's interpretation of the experimental data." Bleier provided many examples of anthropomorphic (based on human qualities) and androcentric (male-centered) biases in both animal and clinical research, such as the following one associated with the study of aggressive behavior in rats: "Observations were made that male rats in a cage fight; female rats do not. When given an electric shock, the male rats fight more; the females do not. While this may be considered proof that males are *naturally* aggressive, what about the equally 'obvious' conclusion that females may be more intelligent, since fighting each other is clearly an ineffectual response to being shocked by some human being?" As to the study of *human* biology and behavior, even more fraught with the risk of personal bias, Gould and Lewontin agreed contemporaneously with Bleier that it is more tenable to argue in terms not of biological determinism but of biological potentiality, since societal factors may affect biological expression.[16]

One final point regarding contemporary influences on scientific discourse must not escape our attention: scientific texts and their language also may be subject to the pressures or biases exerted by conflicts of interest in the corporate world. Corporate researchers must answer to their profit-minded employers. Given the inherent secrecy that such a competitive science-for-profit environment fosters, it is not too hard to imagine how the language and wording of corporate scientific documents could run a higher than ordinary risk of being scientifically questionable. We need only recall how many years and legal battles it took for the truth to finally surface from the reports of tobacco company scientists on the addictive and carcinogenic qualities of their products, eventually leading to cautionary language on the packaging itself.[17] Are similar contests brewing over the accuracy of scientific language associated with synthetic nutritional supplements or substitutes, or crops and animals that are genetically modified? These examples of the sociopolitical contexts that exert a shaping influence on scientific texts are intended only to underscore the reality that science is written by people, and consequently it contains all the potential for glory and failure that humans encompass.

SCIENTIFIC ENGLISH IN ACTION

An understanding of the historical evolution, philosophical orientation, and practical ethic of scientific English provides the contexts needed to grasp its principles for sound practice. Using scientific English to communicate plainly and readably requires certain compositional strategies, from the level of words and phrases to that of sentences and paragraphs, which will be illustrated in the remaining sections of this chapter. Upon deeper consideration of the cliché that the facts (or data) speak for themselves, meticulous and experienced users of scientific English will realize that this is not so. It is the writer who must fashion a thesis, gather and evaluate information, make conclusions, and then find the best scientific English to communicate it all as accurately as humanly possible in a coherent account that both enlightens and convinces the reader. Michael Katz asserts that scientific prose must build a narrative that is readable and that has a "smooth, flowing style [with] balanced and cogent wording." For Katz, the essence of an effective scientific style is in constructing crystal clear sentences: "Each sentence must convey a definite idea, and it must have an unequivocal interpretation: there can be no mystery, no vagary, and no intimations of unwritten meanings or of arcane knowledge." On the other hand, this view must be tempered by that of George Gopen and Judith Swan, who agree with Katz but also remind us how difficult it is to achieve complete and unequivocal clarity and objectivity in language. Gopen and Swan argue that "we cannot succeed in making even a single sentence mean one and only one thing; we can only increase the odds that a large majority of readers will tend to interpret our discourse according to our intentions." The researcher-writer's challenge is to try to ensure that the reader will readily decode virtually the identical meaning that the writer intended to encode and transmit. To achieve this rigorous standard, Katz advises scientists to "use simple, direct words, words with little emotional weight and clear meanings."[18]

The specific examples and strategies offered here are intended to serve two interrelated purposes: first, to illustrate some basic principles of usage in scientific English, and second, to provide practical guidance in making choices that favor maximum plainness in scientific prose. Morris Freedman pointed out what he called the seven sins of technical writing, all of which apply to scientific English, with the primary one being an "indifference" that neglects the reader. From this act of neglect follow the six other transgressions, which he terms fuzziness, emptiness, wordiness, bad habits, deadly passive, and mechanical errors. In the remaining pages of this chapter, most of these hazards are addressed in some form. Making the best choices presupposes a critical writer who is mindful of the expectations of the reader. Some of the examples of scientific English in action are quoted from actual use in scientific research articles. It will become apparent that length and complexity of the various examples range from simple words and phrases to sentences and paragraphs of varying sophistication and complexity. The examples are grouped for convenience into these primary areas: objectivity and precision; clarity and coherence; simplicity and conciseness; misused words and phrases; and punctuation.[19]

OBJECTIVITY AND PRECISION

Objective and precise scientific English is obtained through a range of practices, including making congruent pronoun references; using passive versus active wording; using tense precisely; using concrete versus abstract wording; denoting versus connoting; using numerical expression; articulating action and narrative focus; ensuring logical continuity; and avoiding unnecessary, useless, and dense language. Being precise and objective in scientific writing means choosing words for their accuracy, specificity, and concrete materiality. To the researcher, objectivity also means downplaying the human writer by avoiding words that have personal or emotional values and references, focusing instead on Baconian *things.*

Although quantification of observations and findings is the most easily recognizable form of objective and precise scientific English, there is more to it than meets the eye. Quantification itself must be tested against the logic, rationale, and range of human interpretation that underlie, surround, and support scientific statements. Numerical support must be accompanied by precise and objective wording as well, and this means the writer must be concerned with issues as simple as the use of first-person singular pronouns.

PRONOUN REFERENCES

Other than when referring to one another's research or to clinical cases, as writers scientists tend to avoid making references to human beings, especially to themselves in the first person. Katz advises, however, that a writer should "not be afraid of using first-person singular pronouns when they are appropriate. 'I propose' (or 'we propose') is better than 'it is proposed.' For single-author papers, do not use 'we' or 'our,' unless you are actually referring to things shared by others." Use of "I," "me," "my," "mine," or "our" does not automatically threaten the precision or objectivity of a scientific statement. In the now famous paper announcing their elucidation of DNA's structure, Watson and Crick began with a succinct declarative sentence that is self-referential yet reserved (given their discovery's magnitude).

> Ex. 1.1
> We wish to suggest a structure for the salt of deoxyribose nucleic acid
> (D.N.A.).

Similarly, a sentence can begin with "our" and still convey the same clarity
and precision that Watson and Crick's does.

> Ex. 1.2
> Our data show that most (51.5%) of the subjects were inaccurate in their es-
> timation of the number of episodes of nocturia per night.

Those wishing to avoid the personal references often make substitutions like
"this paper suggests" for "we wish to suggest" (Ex. 1.1), or "the data in this
study show" for "our data show" (Ex. 1.2), though some constructions are
wordier. Such simple differences are merely personal stylistic choices and do
not affect the information's scientific meaning.[20]

PASSIVE VERSUS ACTIVE WORDING

To avoid personal references, and for other reasons, researchers use passive
wording regularly. Some may see passive constructions as weak, as giving a
specious objectivity to research, and as omitting human agency to avoid accountability. Though these criticisms are not without merit, passive wording
sometimes is either unavoidable or beneficial. We can therefore note instances
of both appropriate and counterproductive uses of passive wording.

In the following sentence, there is no direct human agency (the relevant
word choices are emphasized here):

> Ex. 1.3
> One molecule of ethanol *is first metabolized* to the very toxic compound ac-
> etaldehyde, which in turn *is rapidly catabolized* by at least six different en-
> zymes to yield acetate, which *is converted* to acetyl coenzyme *A* to enter the
> Tricarboxylic Acid Cycle and finally yield the end products of carbon diox-
> ide and water.

Even when there is a person as direct agent, readers of scientific papers normally focus not on the agent but on the research. For instance, passive wording is common in procedural descriptions such as this one:

> Ex. 1.4
>
> Preference testing was carried out . . . Sixty naïve mice from each strain were tested . . . Each animal was housed . . . Measurements of the amount of fluid consumed from each tube were taken . . . The position of the tubes was alternated daily . . . The preference index was derived . . .

Revising the passives in Ex. 1.4 to provide an agent makes the subject "we"—that is, we carried out, we tested, we housed, we measured, we alternated, we derived—instead of preference testing, mice, animal, measurements, tubes, and preference index. Such a revision would make the description more direct and the scientist more accountable, but the subject will have shifted from the objects to the author, who is of little interest relative to the procedure described. Moreover, the new subject "we" and its verb at the beginning of each sentence are barriers in front of the material of real interest.

Besides keeping the information (versus the writer) at center stage, passive constructions also permit more nuanced wording with a more precise focus through relative emphasis of sentence content. Consider the different placements of emphasis in this pair of sentences with the same content:

> Ex. 1.5
> 1. A jaw-jerk response was elicited quite strongly and visibly by an intraperitoneal infusion of 10% ethanol.
> 2. An intraperitoneal infusion of 10% ethanol elicited a jaw-jerk response quite strongly and visibly.

The first sentence emphasizes jaw-jerk elicitation, while the second puts more emphasis on the intraperitoneal infusion. The writer must decide which option works best for the desired denotation in a particular scientific context.

There are also uses of passive or active wording that weaken the rigor of scientific English, as in the inconsistent wording here:

> Ex. 1.6
> In order to investigate the NMR line broadening in more detail (incomplete reaction can also give rise to such broadening), *we performed* spin-lattice (T_1) and spin-spin (T_2) relaxation measurements. The results for some selected atoms of the N-tBOC-L-phenyl-modified dendrimers (generations 1 to 5) *are given* in Fig. 2.[21]

Where the sentences in Ex. 1.5 illustrate readability challenges that impede smooth flow, our concern in Ex. 1.6 is the mixed use of wording that is active ("we performed") and passive ("are given") when there is consistent human agency. For the two sentences to have a parallel voice, the second sentence could be revised to begin actively: "Fig. 2 shows the results . . ."

The following sentence is weakened by unnecessary indirectness:

Ex. 1.7

The increments in open-field time needed to be small enough so that the new outdoor environment could be readily adapted to by the animals.

It should be revised to read actively and directly: "so that the animals could readily adapt to their new outdoor environment." In these sentences, the same point is stated with progressively more direct wording:

Ex. 1.8

1. Conversion from manual to automated measurement was effected.
2. Manual measurement was converted to automated measurement.
3. [We, they, the industry] converted from manual to automated measure-ment.

The different versions in this example have a somewhat different locus of emphasis, but the notion of conversion is unaltered.

Use of the passive in itself does not confer objectivity or greater precision and sometimes may instead weaken flow and readability. In any particular context, writers must decide whether to use a passive or active construction in relation to such factors as agency, logic, readability, focus, emphasis, and consistency.

PRECISION IN TENSE USAGE

A good writer also maintains a precise and objective scientific narrative through the appropriate use of verb tense in different parts of a document. Verb tenses are an important means of differentiating between the reporting of experimental observations (performed in the past) and their discussion (which includes present commentary). The writer should not generalize and report the findings from an experiment in the present tense, as though they were universal or general truths. Consider the tense options in these sentences:

Ex. 1.9

1. Smith and Jones (2002) found [versus *find*] a sharp decrease in serotonin at day 4.
2. We detected [versus *there is*] a sharp decrease in serotonin at day 4.

In the first sentence, replacing the past tense by the present tense ("find") leads to an inaccurate statement, since Smith and Jones are not continuing their experiment and it is not certain that the same result would occur if they did. In the second sentence, using the present tense ("there is") changes the statement of a research finding into one of generally accepted knowledge.

Clarity is also compromised when a writer uses the present tense to express a prior finding in a way that indicates its continued truth in the present, as follows:

Ex. 1.10

It *was found* that the level of acetaldehyde in the blood *increases* [versus *increased*] with chronic alcohol consumption.

Such a result might not be the case in future studies. Therefore, reporting the result entirely in the past tense (i.e., "increased") is not only more accurate but also ensures that this statement would remain accurate in the future, even with different findings. Writing the statement entirely in the present tense— "Chronic alcohol consumption increases blood acetaldehyde levels"—generalizes inaccurately.

The same decisions about tense must be made in a report's conclusion. Although the present tense lends itself to general discussion, tense in concluding statements based on the findings must be carefully considered, as in this example:

Ex. 1.11

We conclude that carbohydrate loading affected [versus *affects*] endurance.

Using "affected" maintains the conclusion as a past inference of the past results, while using the present tense ("affects") creates a general statement regarding the results, which is scientifically less accurate. Conclusions about the research may be in the present tense, but those that generalize a finding should be kept in the past tense:

Ex. 1.12

The close correspondence between the chemical uptake by plants and the RWD indicates [versus *indicated*] that the rate of root growth *was* more important than the specific absorption rate.

Statements in the past tense generally are more rigorous scientifically. The regular use of present tense is more appropriate in discussions or in developments that include mathematical equations.

CONCRETE VERSUS ABSTRACT WORDING

Scientific expression relies for its accuracy and objectivity on concrete and specific senses in its language. Expressions that are abstract or vague are of little use because they contain very limited and imprecise information. Information expressed concretely can be decoded through our five senses and is more useful scientifically. Consider the difference in the level of precision and detail between these two sentences:

Ex. 1.13

1. Researchers have found that experiments with crops under reduced lighting require a considerable amount of time because the seeds germinate so slowly.
2. Johnson and Brown (2003) have found that experiments with tomatoes and carrots in 50% and 75% light-deprived environments require 12–16 weeks instead of 7–8 weeks because the seeds take twice as long to germinate.

Or, note how markedly the following two versions of the same observation differ in the specificity with which they express information based on sight and sound:

Ex. 1.14

1. The animal was multicolored and made an annoying sound.
2. The animal was brown with white spots widely and evenly distributed over its fur and it growled sharply, loudly, and continuously like a menacing bulldog.

Mindful of purpose, audience, and context (i.e., surrounding sentences), writers must use the appropriate level of detail and specificity in their technical descriptions.

DENOTATION VERSUS CONNOTATION

Given that scientific language works through a progressive narrowing of reference, terminology, and meaning, researchers require language that has pinpoint precision, concrete and specific senses, and very limited connotative expression. To the extent humanly and professionally possible, scientific writers must denote their ideas and results precisely and unambiguously. Impediments to precise denotation include general, vague, or abstract words, indeterminate or inaccurate references, poorly chosen figures of speech, and anthropomorphic language. Scientific words and statements should "denote," or stand for literal and unequivocal meanings, rather than "connote," or suggest associated meanings that would cloud their objectivity, be imprecise, and undermine their utility. Using language that is anthropomorphic, pretentious, or intended to inject humor, or words that have a range of colloquial senses, is not consonant with scientific denotation. For instance, a word like "adequate" actually may have a negative connotation rather than the intended denotation of "sufficient for what is needed." Would an employer be tempted to hire a job applicant who writes in the cover letter that his or her qualifications are merely "adequate" for the job? Would you stand on a construction scaffold that has merely "adequate" support? These connotations can actually imply *limited* capacity or safety. Is an animal that displays force against an approaching human behaving "nastily"—unjustifiably "attacking"—or, more objectively, is it simply being "aggressive" in order to defend its territory?

Anthropomorphism

One dangerous type of connoting is using language that is anthropomorphic, conferring human agency, intent, or qualities on a nonhuman entity. Using language anthropomorphically in scientific documents can work to undermine the authority of a writer and the validity and reliability of the information. Consider the anthropomorphic wording in this sentence:

Ex. 1.15

The C57 strain of mice *liked* [or *preferred,* versus *selected* or *drank*] ethanol more than butanediol by a factor of ten.

The researcher knows from the consumption data that the mice chose to drink ethanol over butanediol, but cannot know whether they actually like or prefer (connoting *enjoy* or *seek*) either of the compounds for their qualities of taste or pharmacologic effect.

The following sentence contains another form of anthropomorphism:

Ex. 1.16
Alcohol drinking research has largely ignored the different neurological symptoms of alcohol abuse and typically has been content to view them narrowly as secondary effects of heightened neurotransmitter release.

The subject, alcohol drinking research, is treated as a purpose-driven agent that is free to "ignore" things or be "content." Avoid such anthropomorphic constructions by providing an agent or focusing on the research information, as in the following passive and active options:

Ex. 1.17
1. The different neurological symptoms in alcohol abuse patients have largely been ignored in alcohol drinking research, and narrowly viewed by most investigators as secondary effects of heightened neurotransmitter release.
2. Investigators have largely ignored the different neurological symptoms in alcohol abuse patients, and narrowly viewed them as secondary effects of heightened neurotransmitter release.

There are also anthropomorphic statements that are teleological, ascribing to nonhuman entities the intention to do something. In the following pair of sentences, the teleology in the first option is corrected in the second option:

Ex. 1.18
1. The chromatography *columns packed* their gel differently.
2. The packing in the chromatography columns differed.

In this case provided by the APA *Publication Manual,* the anthropomorphic and gender-biased analogy implicit in the first sentence is corrected in the second:

Ex. 1.19
1. Ancestral horses probably traveled as wild horses do today, either in bands of bachelor males or in harems of mares headed by a single stallion.
2. Ancestral horses probably traveled as wild horses do today, either in bands of males or in groups of several mares and a stallion.[22]

References to Humans

Another area of scientific English that calls for objective wording is that of references to humans more broadly. Research documents must be devoid of language that is biased or otherwise subjective when referring to such features as sex, age, and disability. The inconsistent gender references emphasized in the first version of this procedural description are corrected in the second sentence by rewording it to avoid the restrictive pronouns:

Ex. 1.20

1. We used a *mixed-sex* sample and screened each volunteer for *his* prior exposure to the toxins with a questionnaire and fluid samples that *he* submitted at the first office visit.
2. We used a mixed-sex sample and screened each volunteer for prior exposure to the toxins with a questionnaire and fluid samples that *we* collected at the first office visit.

Other expressions that permit precision and objectivity are "his or her" and "he or she" in place of single-sex references (although many people frown on "s/he" and "he/she"). Some writers prefer to alternate the sex-specific pronouns throughout a text, saying first "he" or "his," followed by "she" or "her" the next time, although this solution can sometimes be confusing for readers. A common form of imprecise reference is the use of single-sex words like "man" or "mankind"; more neutral words like "humankind," "humanity," "human beings," "humans," or "people" are more precise and avoid the potential for alienating some readers.

At the same time, however, maintaining linguistic objectivity should not result in dehumanizing language, such as referring to people as "subjects" or "cases" in clinical studies (Ex. 1.2); a writer can use more positive and more accurate options like "participants," "respondents," or "clients." Likewise, instead of referring to a person who has a particular medical condition as one who "is afflicted with," "is a victim of," or "suffers from" that condition, write that the person "is myopic," for instance, or "is a cancer patient," or "lives with" the condition. Similarly, when referring to persons with disabilities choose neutral phrasing like "uses a wheelchair" rather than "is wheelchair-bound," or "wears [versus *is dependent on*] a prosthesis." References to people of advanced age should avoid catchy or imprecise words like "seniors," "golden-agers," or "older people" in favor of more specific wording like "septuagenarians" or "seventy-year-old retirees." Objective *and* humane refer-ences to people keep the focus on the information rather than on the biased ex-pressions of a given writer or culture.[23]

Figures of Speech

A form of connotative language that is emblematic of the subjective and im-precise communication that scientists resist, but nonetheless use in important ways, is the figure of speech. Metaphors, similes, analogies, and other such literary expressions are regarded today, no less than in Bacon's time, as a threat to scientific precision and objectivity. The biologist Antoinette Wilkin-son put it this way: "The figure of speech may magnify, diminish, emphasize, heighten, or color the idea expressed, or it may cast a particular light on it or give it a particular tone" that scientific writing should not have.[24] This is espe-cially so of metaphors or analogies that are highly colloquial. In this example, the improper use of a common metaphor in the first sentence is revised in the second version for more precise denotation:

Ex. 1.21
1. A hamster being placed in a cage already housing a mouse is *like a train wreck waiting to happen.*
2. Placing a hamster in a cage already housing a mouse will trigger a relent-less territorial aggression in the mouse, which—due to the animals' size and strength differences—is ultimately lethal to itself.

The train-wreck figure, while certainly powerful in ordinary everyday lan-guage, represents the kind of informal or colloquial comparisons that under-mine the objectivity and precision of scientific English.

However, used with appropriate professional restraint, figures of speech can enhance the clarity of scientific information or ideas. They can provide a particular and compelling exactitude of either image or thought. Consider the following simple example (emphasized) from a highly technical article on DNA's structure:

Ex. 1.22
All the bases are flat, and since they are stacked roughly one above another *like a pile of pennies,* it makes no difference which pair is neighbor to which.

Or, here is a sophisticated example of a metaphor used in a paper by Stephen Jay Gould and Richard Lewontin on evolutionary adaptation:

> Ex. 1.23
> We strongly suspect that Aztec cannibalism was *an "adaptation" much like evangelists and rivers in spandrels, or ornamented bosses in ceiling spaces:* a secondary epiphenomenon representing a fruitful use of available parts, not a cause of the entire system.

The power of figurative language as an aid in both making and communicating scientific syntheses was discussed earlier in this chapter, as in the cases of "the double helix" and genetic "coding." A memorable metaphor used in the 1970s by the biologist Lewis Thomas in his writing for a general audience is that of the earth as a "single cell." It is now commonplace to refer to computer "viruses" or to "smart" technology, making analogies to the human associations with these terms. Figurative wording is also useful when scientists wish to explain an idea to readers in the general public. In a *Scientific American* article on the role of glial cells in neuronal stimulation (or "firing"), for instance, a biologist describes one experimental approach with the following analogy.

> Ex. 1.24
> Using a sharp microelectrode, they cut a line through a layer of astrocytes in culture, forming a cell-free void that would act like a highway separating burning forests on either side.[25]

The point that requires emphasis here is that such figurative language, as risky as it is for exactitude in scientific exposition, can indeed be used beneficially by researchers within the limited range of their discipline's linguistic freedom, just as (recalling Luria's apt analogy) a virtuoso violinist has a limited degree of freedom in interpreting a musical score.[26]

Other practices that are not consonant with denotative language involve wording that draws attention to the writer or a stilted expression, such as in the use of seemingly pretentious words—for example, aforementioned, commence, thereof—or even the use of humor to lighten the formality of scientific writing (sometimes an urge especially at conference presentations). To an interested reader or listener such practices are unnecessary, intrusive, and out of character with the declarative, documentary, and objective nature of scientific writing. The attempt at humor in the following sentence is not only blatantly sex-biased but gravely undermines scientific objectivity and precision and has no place in scientific English.

Ex. 1.25

It has been observed that the female praying mantis consumes the male's head after mating, *but I'll refrain from the tempting analogies to humans.*

Moreover, such writer-centered language, whether stilted or light-hearted, may also be a source of confusion and annoyance to international audiences. Medawar underscores metaphorically the importance of denotative clarity and reader-centeredness in scientific writing: "A good writer never makes one feel as if one were wading through mud or picking one's way with bare feet through broken glass."[27] It is never a good idea to take your readers for granted by neglecting their needs and expectations.

NUMERICAL EXPRESSION

"A number," states the style manual of the Council of Biology Editors (CBE), "is the representation in numeric or word form of a count, an enumeration, or a measurement."[28] Given how critical measuring and quantifying are to researchers, much could be said about how they use and express numbers, including mathematical symbols and equations (just consider the use of formulaic representation in population genetics or in enzyme kinetics). Numerical expression in scientific prose must be accurate, unambiguous, and consistent. Our concern here is with the simplest and most common numerical references or usages, for which writers are nonetheless error-prone. Examples of these are numerical agreement, spelling out, ranges, and hyphens and dashes.

Number Agreement

A common and sometimes implicit (so almost unnoticed) form of numerical expression involves number *agreement,* as in each of the following cases:

Ex. 1.26

1. Our *data* suggest [not *suggests*] a wide intraspecies variation.
2. The non-drinker *rat* avoids EtOH after its [not *their*] first exposure.
3. Ninety-five percent of the *animals* are [not *is*] tested annually.
4. A *clan size* of 15 to 20 is [not *are*] occasionally observed.
5. Six to eight is typical, but *10–12* mice per litter *is* [not *are*] common.

Note that "data" is plural for "datum" and should be used as such. In the other cases, a careful look at the referents will reveal the appropriate and complementary options.

Spelling Out Numbers

Numbers and percentages typically are spelled out either at the beginning of a sentence or to reduce the risk of ambiguity. They are not spelled out when used with unit abbreviations or symbols. The following sentences illustrate such practices:

Ex. 1.27
1. Eighty-five percent [*not* 85%] of the animals usually survive, but there is a 60% to 95% [or 60%-95%] range.
2. The animals consumed 10 five-pound [*not* 5-pound] bags of feed.
3. We discovered five [*not* 5] 500-year-old artifacts.
4. The animals drank 20 ml [*not* twenty ml] of water.

When adjacent numbers are not given contrast by spelling out as needed, readers are unnecessarily exposed to ambiguity and imprecision. Without the noted distinctions in numerical expression, hurried readers of the second sentence above could mistake *ten individual units* weighing five pounds each (a total of 50 pounds) for an unspecified number of bags each weighing 105 pounds, a considerable difference. In the third sentence, the reference to *five* artifacts each dated 500 years old could be mistaken for an indeterminate number of artifacts each dating 5,500 years.

Numerical Ranges

Among the elements to consider in expressing numerical ranges are: when to spell them out; how to use dashes or symbols (e.g., %, or Hz) with them; distinguishing year, page, and mixed number-symbol ranges; and whether to interpose "to" or "through" between the items. These elements are illustrated in the following set of sentences, with brackets and emphases to indicate the various options.

Ex. 1.28
1. We checked the government data for 1998 *through* 2003 [or 1998–2003, not 1998–03 or 1998 to 2003], which were tabulated over pages 146 *to* 149 [or 146–49] of the document.

---

2. The data are from 20 of the animals, which we labeled "DBA1278" to "DBA1298" [*but* "DBA 1278–98," with letters and numbers separated].
3. Twenty-four percent to twenty-nine percent [*not* 29%] of the animals typically are non-drinkers, but their range in inter-strain selection is 12% to 89% [or 12%-89%; *not* 12–89%]. Their weight range was 2 to 5 lbs [or 2–5 lbs, *not* 2 lbs to 5 lbs].

In the first sentence, "through" is used to indicate that 2003 is included, versus "to" 2003. In the third sentence, the twenty-nine is spelled out to be parallel in form with the spelled-out number that begins the sentence.

Using Hyphens with Numbers

The use of hyphens in numerical references or series works both to reduce repetition, thereby streamlining a text and permitting smoother flow, as well as to denote with precision.

Ex. 1.29
1. We checked at 15-, 30-, and 45-minute intervals and observed a threefold [or three-fold; *not* 3-fold] increase in metabolic rate.
2. We discovered 200-year-old [versus 200 year-old] trees.

CLARITY AND COHERENCE

Even after the researcher-writer has achieved the requisite objectivity and precision in a document's language, the work is not done. The document must also read clearly and coherently so that its scientific content is fully decoded and thereby usable by readers. Scientific documents are not among the easiest texts to read. Science is supposed to be that way, goes the common view, with all the detailed facts and special terminology that scientists use to describe what they do and observe and to explain complex ideas. There is, however, still no good excuse for the fuzziness in the first of these sentences, which is translated in the second version:

Ex. 1.30
1. When the element numbered one is brought into tactual contact with the element numbered two, when the appropriate conditions of temperature have

> been met above the previously determined safety point, then there will be exhibited a tendency for the appropriate circuit to be closed and consequently to serve the purpose of activating an audible warning device.
>
> 2. When the heat rises above the set safety point, element one touches element two, closing a circuit and setting off a bell.

George Gopen and Judith Swan, in their influential essay "The Science of Scientific Writing," argue that "complexity of thought need not lead to impenetrability of expression." Scientist-writers can achieve clarity, without having to oversimplify, when they fulfill basic reader expectations. The following sections summarize the writing method Gopen and Swan recommend in their essay describing these key areas of reader expectation, which include: articulation of action, narrative focus, relative emphasis of content, and logical continuity.[29]

ARTICULATION OF ACTION AND NARRATIVE FOCUS

Scientific writing depends on action words for constructing a sensible narrative to describe or explain what occurred and how, when, where, and why it occurred. This ranges from what scientists themselves do procedurally—setting up, conducting, measuring—to what they observe happening in the course of their research. Therefore, it is by necessity that verbs, especially technical ones (e.g., catherized, autoclaved, denucleated), are extraordinarily abundant in scientists' writing. Because of this pervasive articulation of action, readers will be grateful for wording that facilitates rather than obstructs the action's coherence.

The first important gauge, according to Gopen and Swan, of whether action is being clearly articulated is the relative positions of verbs and their grammatical subjects. In the most effective constructions, grammatical subjects are followed as soon as possible by their verbs. The second of the following two options is more readable not because of its brevity but rather because it narrows considerably the distance between subject and verb.[30]

> Ex. 1.31
> 1. *The high-ethanol-selecting mice* (C57BL/6j), which were given 10 ml of a high-protein, low-fat solution twice daily over a 60-day pretest period, *consumed significantly less* alcohol than the low-ethanol-selecting (DBA/2J) mice.
> 2. The *high-ethanol-selecting mice* (C57BL/6j) *consumed significantly less* alcohol than the low-ethanol-selecting (DBA/2J) mice.

In the first version, the reader must wait some 20 words before seeing action and learning that the subject—high-ethanol-selecting mice—*consumed significantly less* alcohol than the low-ethanol-selecting strain of mice. The second version, with all the intervening material between subject and verb removed, is more direct. Though the second version is shorter, deciding which version is better in this case has less to do with length than with the writer's purpose and whether the additional contextual information is needed within the same sentence.

As Gopen and Swan point out, even more of an obstacle to clarity and coherence than the *separation* of subject and verb is the challenge of figuring out the *location* of the action and its significance in the overall narrative. In the following paragraph, the emphasized verbs—"is," "are presumed to be," "are transcribed," "has," "can be alleviated," "destabilizes"—are of little help for understanding the point of the writer's narrative.

> Ex. 1.32
> Transcription of the 5S RNA genes in the egg extract *is* TFIIIA-dependent. This *is* surprising, because the concentration of TFIIIA *is* the same as in the oocyte nuclear extract. The other transcription factors and RNA polymerase III *are presumed to be* in excess over available TFIIIA, because tRNA genes *are transcribed* in the egg extract. The addition of egg extract to the oocyte nuclear extract *has* two effects on transcription efficiency. First, there *is* a general inhibition of transcription that *can be alleviated* in part by supplementation with high concentrations of RNA polymerase III. Second, egg extract *destabilizes* transcription complexes formed with oocyte but not somatic 5S RNA genes.[31]

Even with only limited knowledge of the writer's intentions, it is evident that the narrative is undermined by verbs that do not articulate precisely what action is taking place, compounded by an absence of topical cues in the sentences. The reader is challenged to figure out the relative importance of the key players: "egg extract," "TFIIIA," "oocyte extract," "RNA polymerase III," "5S RNA," and "transcription." Gopen and Swan's revised version rearranges the narrative so it focuses on "egg extract," by placing it at the beginning of several sentences, and on the mediating effect of "TFIIIA," by locating it in clear relation to "egg extract," as well as prominent use of the key connecting verbs "limit" and "inhibit":

Ex. 1.33

In the egg extract, the availability of TFIIIA *limits* transcription of the 5S RNA genes. This is surprising because the same concentration of TFIIIA *does not limit* the transcription in the oocyte nuclear extract. In the egg extract, transcription is not limited by RNA polymerase or other factors because transcription of tRNA genes *indicates* that these factors are in excess over available TFIIIA. When added to the nuclear extract, the egg extract *affected* the efficiency of transcription in two ways. First it *inhibited* transcription generally; this inhibition could be alleviated in part by supplementing the mixture with high concentrations of RNA polymerase III. Second, the egg extract *destabilized* transcription complexes formed by oocyte but not by somatic 5S genes.

Readers will now see the connection between "limit" and "inhibit" in relation to the writer's hypotheses: that is, that transcription is limited by a TFIIIA inhibitor present in the egg extract, and that the inhibitor's action is detectable when the egg extract is added to the oocyte extract and the effects on transcription are examined. The second version allows readers to focus their energy on evaluating the validity of the author's hypotheses rather than on deciphering a poorly structured and vaguely articulated narrative.

RELATIVE EMPHASIS OF CONTENT

Gopen and Swan show that another clarifying strategy is to take advantage of natural positions of emphasis. This revision of the first sentence in Ex. 1.31 emphasizes a key action:

Ex. 1.34

The high-ethanol-drinking mice (C57BL/6j) were given 10 ml of a high-protein, low-fat solution twice daily over a 60-day pretest period; *they consumed significantly less* alcohol than the low-ethanol-selecting (DBA/2J) mice.

In this version, the subject ("high-ethanol-drinking mice") is connected more directly with the verb "were," and the significant action—"consumed significantly less"—receives emphasis after an added semicolon. If the writer fails to place the most important content in a position of natural emphasis, whether at the beginning or the end of sentences or paragraphs, the reader may misjudge the relative weight of the information, which may then lead to disrupted closure. The revision in Ex. 1.34 keeps the contextual information at the beginning of the sentence while helping the reader by using a semicolon to separate and to accentuate the result that the normally high-alcohol-drinking mice actually "consumed significantly less" than the normally low-drinking mice.

It is not simply the length of a sentence that determines its degree of clarity. Short sentences can be written as confusingly as long ones. Positions of emphasis used strategically will assist readers in evaluating the relative importance of various pieces of information. Gopen and Swan define an excessively long sentence as one that cannot accommodate all the items requiring stress.

LOGICAL CONTINUITY

Another readability-enhancing principle described by Gopen and Swan involves the strategic placement within sentences and paragraphs of new versus "old" or contextual information, so that the writer can avoid any discontinuities or logical gaps in the presentation of information. The guiding principle here is that rather than rushing to present the new information at the beginning, or "topic" position, of a sentence or paragraph, it is helpful to begin with "old" information that connects backward to provide context. That way, when the new information is given later in the sentence or paragraph—in the "stress" position—reader expectations will not be thwarted by logical gaps. The beginning of a sentence or paragraph is commonly referred to by reading and writing theorists as the "topic" position because that is where it is logical to introduce the controlling idea as well as to provide a transition from or a link to earlier, or "old," information. The end of a sentence or paragraph is commonly viewed as a position of stress because those words are read last and therefore tend to be cognitively emphasized.

Consider the following paragraph, with special attention to the emphasized phrase:

Ex. 1.35
The ability of liver extracts from the C57 and DBA mouse strains to reduce NAD with ethanol and 1,3-butanediol as substrates was measured using an *arbitrarily selected* set of assay conditions. The NAD reduction in extracts was evaluated using the conditions for determining acetaldehyde dehydrogenase activity described by Sheppard (1968). From NAD reduction assays, the specific activities with 1,3-butanediol and ethanol are 13.54 for the C57 strain and 6.84 for the DBA strain.[32]

The paragraph suffers not only from subject-verb separation ("ability . . . was measured"), but its readability problems are compounded by a missing connection, or a logical gap, between pieces of information—that is, no explanation is provided for using the "arbitrarily selected" measurement conditions "described by Sheppard." The following revision improves the logical flow by adding contextual and linguistic connections to permit readers to comprehend fully the researcher's methods and findings.

Ex. 1.36

Assaying liver extracts for dehydrogenase activity with 1,3-butanediol poses problems that make attempts to determine individual dehydrogenase activity no more informative than evaluating the ability of extracts to reduce NAD with 1,3-butanediol as substrate. These complications make it impossible to know the required number of assay stages. Therefore, we measured the ability of liver extracts from the C57 and DBA mouse strains to reduce NAD with ethanol and 1,3-butanediol as substrates using an arbitrarily selected set of assay conditions. To evaluate the NAD reduction, we used the conditions for determining acetaldehyde dehydrogenase activity described by Sheppard (1968). *Since our comparison is between the extracts of two mouse strains with each alcohol and not between the two alcohols themselves, knowledge of the exact number of steps assayed is not essential for meaningful data evaluation.* Our NAD reduction assays with 1,3-butanediol yielded specific activities of 13.54 for the C57 strain and 6.82 for the DBA strain, and with ethanol 1.6 for the C57 strain and 1.94 for the DBA strain.

The two emphasized sentences that now precede the original first sentence, together with the logical connective "therefore," explain the reason for the "arbitrarily selected" assay conditions of Sheppard. A further logical gap is filled by the italicized passage preceding the final sentence validating the numerical results (13.54 and 6.82). Note also that the revised version adds more directness by using active wording ("we measured," "we used"), improves the articulation of action by closing the 19-word subject-verb distance in the original first sentence ("The ability . . . was measured" to "we measured"), and clarifies agency and attribution by adding "our" in the final sentence.

The following highly technical and involved case from Gopen and Swan (Exs. 1.37 and 1.38) illustrates how old and new information can be sequenced and contextualized in a paragraph to maintain logical continuity. First, here is a paragraph that omits important connective information:

Ex. 1.37

The enthalpy of hydrogen bond formation between the nucleoside bases
2′deoxyguanosine (dG) and 2′deoxycytidine (dC) has been determined by
direct measurement. dG and dC were derivatized at the 5′ and 3′ hydroxyls
with triisopropylsilyl groups to obtain solubility of the nucleosides in non-
aqueous solvents and to prevent the ribose hydroxyls from forming hydrogen bonds. From isoperibolic titration measurements, the enthalpy of dC:dG
base pair formation is -6.65 ± 0.32 kcal/mol.[33]

There are various readability problems in this paragraph that go beyond the
technical difficulty associated with its specialized terminology. The problems
that interfere with the narrative's accessibility to the reader include: an uncertain main player, subject-verb separation, poor use of stress positions, and
missing contextual information that impedes logical clarity. Gopen and Swan's
revision of the paragraph addresses the problems and results in a sequence of
improved dynamics.

Ex. 1.38

We have directly measured the enthalpy of hydrogen bond formation between the nucleoside bases 2′deoxyguanosine (dG) and 2′deoxycytidine
(dC). dG and dC were derivatized at the 5′ and 3′ hydroxyls with triisopropylsilyl groups; these groups serve both to solubilize the nucleosides in
non-aqueous solvents and to prevent the ribose hydroxyls from forming hydrogen bonds. Consequently, when the derivatized nucleosides are dissolved
in non-aqueous solvents, hydrogen bonds form almost exclusively between
the bases. Since the interbase hydrogen bonds are the only bonds to form
upon mixing, their enthalpy of formation can be determined directly by measuring the enthalpy of mixing. From our isoperibolic titration measurements,
the enthalpy of dC:dG base pair formation is -6.65 ± 0.32 kcal/mol.

First, the more direct beginning narrows the subject-verb distance and places
"dG" and "dC" in a position of emphasis as "new" information. Second, in the
next sentence dG and dC are kept in the "topic" position as now "old" or familiar information, and "triisopropylsilyl groups" are in the stress position as
new information. Third, the added semicolon in that second sentence sepa-
rates a clause that allows "these [triisopropylsilyl] groups" to be in the topic
position as new information whose important effects are then described.

Fourth, since there are two such important effects, the added "both" alerts the reader to these coming effects, which are in the stress position as two new items of information. Fifth, the added third and fourth sentences—beginning with "consequently" and "since," respectively—fill in critical gaps that now permit logical continuity. Those two inserted sentences explain how the "derivation" mentioned earlier (in the second sentence) relates to the "titration measurements" given in the final sentence. Sixth, in the last sentence, "measurements" now links back to the phrase "determined [measured] directly" in the preceding sentence. Finally, Gopen and Swan's revised paragraph has a logical unity and flow from sentence to sentence, including the linkage or fulfilling symmetry of "measurements" in the final sentence with the "we have directly measured" that began the paragraph.

SIMPLICITY AND CONCISENESS

The physicist Michael Alley notes that "conciseness follows from pursuing two other language goals: being clear and being forthright." To be forthright one must also make scientific statements as simple as possible. The simplest statements, then, are also likely to be the most concise. As Wilkinson asserts, "writing is concise when everything that needs to be said is stated in whatever detail is needed in as few words as possible—that is, just the right words and the right number of words, in the right order—no more, no less." To be simple and concise, scientists can avail themselves of various word-sparing strategies that also work to enhance directness. Such strategies involve reducing verbiage by avoiding redundancy, circumlocution, and useless words. Simplicity can also be maintained by avoiding long strings of technical nouns and adjectives to create phrases that are practically impenetrable, requiring re-reading that wastes the reader's time.[34]

REDUNDANCY AND REPETITION

Redundancy and repetition may occur either within sentences or in different parts of a document. While repeating or rephrasing key information can be helpful in books for emphasizing certain ideas recursively, for making connections, or for facilitating transitions, in shorter documents such as scientific papers it wastes the reader's time. Note the unneeded repetitive wording here:

> Ex. 1.39
> With our open-field activity apparatus, we found that the animals lost 25% *of their motor activity* after 2 hours, 60% *of their motor activity* after 4 hours, and 90% *of their motor activity* after 6 hours.

In the following pair of sentences, the redundant phrases in the first version have been trimmed for a simpler second version.

> Ex. 1.40
> 1. At 10 months *of age,* the animals received *repeated* daily injections of *a percentage* of 3.4% cycloheximide.
> 2. At 10 months, the animals received daily injections of 3.4% cycloheximide.

There are too many variations on this theme to list, but some examples of redundant phrasing are:

- many *in number*
- blue *in color*
- hydroxylation *reaction*
- conical *in shape*
- eliminate *completely*
- the reason *why*
- small *in size*
- at this *moment* [or *point*] *in* time
- scrutinize *closely*
- exact *same*

The emphasized modifiers in the examples restate uselessly the information that is already contained in the modified word.

Redundancy also occurs when the same information is given in different parts of a sentence but with different wording. The following sentence makes the same point twice (once positively, then negatively):

> Ex. 1.41
> Central nervous system sensitivity to propylene differed significantly between the hamsters and rabbits tested in our sample; the two species' sensitivity cannot be said to be anywhere close to identical.

Likewise, stating an idea or point in one part of a paper, such as the introduction—as in the first sentence in the following example—makes it unnecessary to reiterate it in different words later, as the second sentence here does in the discussion section:

Ex. 1.42
1. The concluding section examines three competing theories of male macaques' agonistic behavior during mating. [introduction]
2. Here we will consider three possible explanations for aggressive behavior of males in the mating ritual of macaques. [discussion]

In other cases, writers may state a result in the results section and repeat it soon thereafter in the discussion section. It is well to be mindful that repeating the same information with different wording or sentence structure will not change its essential meaning. The risk of such unnecessary repetition may be higher in the social sciences than in the natural and exact sciences because research papers in the social sciences have more extensive discussion and use terminology that is closer to everyday language.

CIRCUMLOCUTION AND USELESS WORDS

Verbiage in a sentence also results from circumlocution, or roundabout wording that states a point or expresses an idea with too many words. Consider the streamlining from the first to the second version in these sentences:

Ex. 1.43
1. Once this procedure was completed, we proceeded to undertake an investigation of the change in GABA release during the time that amphetamine was not administered.
2. Then we studied the change in GABA release when amphetamine was not administered.

The dozen words that were trimmed from the first sentence are not needed to convey the intended meaning, and the more concise version is of course simpler to read. Overuse of adjectives or noun modifiers also contributes to roundabout wording, as shown by Alley in these two versions of the same idea:

40

Ex. 1.44

1. The objective of our work is to obtain data that can be used in conjunction with a comprehensive chemical kinetics modeling study to generate a detailed understanding of the fundamental chemical processes that lead to engine knock.
2. Our goal is to obtain experimental data that can be used with a chemical kinetics model to explain the chemical processes that lead to engine knock.[35]

As with redundancies, the possibilities for circumlocution are virtually endless, but here are some common examples of roundabout phrasing (with parenthetical revisions):

- in light of the fact that (*because*)
- are in agreement with (*agree*)
- conduct an investigation into (*investigate*)
- it is apparent therefore that (*apparently*)
- of a reversible nature (*reversible*)
- make an adjustment to (*adjust*)
- on two separate occasions (*twice*)
- take into consideration (*consider*)
- has the potential to (*can*)
- in the event that (*if*)
- in close proximity to (*near*)

Authors often use superfluous phrases at the beginning of a sentence. In the following example, the second sentence omits the useless words (emphasized) in the first version:

Ex. 1.45

1. *We conclude here with a summary of the evidence suggesting that,* given certain conditions, disrupted nest-building in mice may be caused by high levels of exogenous testosterone.
2. Our results suggest that high levels of exogenous testosterone disrupt nest building in mice.

The first version begins with words that add nothing to the facts being stated. Moreover, the sentence is weakened further by the qualifiers "may be" and by (unspecified) "certain conditions." A few other parallel examples of useless words to be avoided at the beginning of a sentence are:

- It is interesting to note that
- It is considered that
- It is possible that the cause is
- It is expected that
- It is generally believed that

Circumlocution and useless words make their way into a document for various reasons. Writers may feel that using more words constitutes more thorough explanation. Or some may tend toward affectation and avoid simplicity in favor of elaborate sentences that they believe will sound more scientifically learned. For instance, writers frequently will opt for "utilize" or "employ" when they simply mean "use"; in the noun form, consider the unnecessary phrasing, "The *utilization* [versus *use*] of that technique does not suit our purpose." Yet another reason may be the failure to consider the difference between writing and speech, for writing does not require restatement, repetition for emphasis, or a conversational verbosity typically used when speaking. Whatever the reasons, close attention to how many words are truly needed to convey information will lead to simpler, less convoluted, and more forthright sentences.

NOUN AND ADJECTIVE CLUSTERING

A practice that undermines simplicity in scientific prose is the excessive clustering or stacking of nouns and adjectives. It is common and reasonable to use two-word combinations either of nouns alone or of an adjective with the noun it modifies, such as "agglutinated cells" (adjective and noun) or "heart chamber" (two nouns). When a writer goes beyond such simple combinations, however, the writing begins to lose readability. In the following case provided by Janice Matthews and associates, the first sentence is better off being split into two or more sentences, as in the second version:

Ex. 1.46
1. Five two week old single comb white leghorn specific pathogen free chickens were inoculated with approximately 105 tissue culture infected doses of duck adenovirus.
2. Our sample was composed of white leghorn chickens of the single-comb

---

> variety that were free of specific pathogen. All the chickens in the sample were two weeks old. They were inoculated with approximately 105 doses of tissue culture that was infected with duck adenovirus.

Medawar offers this example of a stacked phrase: "vegetable oil polyunsaturated fatty acid guinea pig delayed type hypersensitivity reaction properties." He suggests that writers may feel encouraged to create such long strings of nouns and adjectives to achieve conciseness that will satisfy the length restrictions set by most periodicals. Nonetheless, the practice of using lengthy clusters of nouns and adjectives illustrates a confusion of conciseness with brevity. Conciseness means using just the *right* number of words, not the *least* number. A writer overly concerned with brevity tends to produce a telegraphic style of writing, relying on linguistic constructions that thwart rather than facilitate simple and clear communication.[36]

MISUSED WORDS AND PHRASES

Writers may misuse some common words due to either inattention or uncertainty about their precise meaning. Words that are susceptible to this problem often come in pairs, like *affect* and *effect* or *that* and *which,* and can be readily confused. In scientific writing, however, such words have specific denotative values. In some cases, the options are closely related in meaning, as in *comprise* and *compose* or *imply* and *infer,* but in other instances the words have entirely different senses, such as *complementary* versus *complimentary.* Using the wrong word in cases like this will undermine the clarity, accuracy, and ultimately the integrity and truthfulness of scientific statements. The following are twenty of the most common pitfalls, accompanied by brief explanations and examples that differentiate among the choices in question. The distinctions are selective rather than exhaustive and focus on scientific uses.

AFFECT AND EFFECT

To "affect" something is to act or to serve in such a way as to cause or produce some outcome or "effect." One may either *affect* (influence, change) or *effect* (bring about, cause) something. The following pair of sentences demonstrates these distinctions.

Ex. 1.47
1. We *affected* their diurnal pattern and produced sleep-deprivation *effects* on their short-term memory.
2. We *effected* a significant reduction in their sleep time, thereby *affecting* their short-term memory.

AFTER AND FOLLOWING

"After" simply means later unless used in the context of a precise time frame. "Following" connotes *immediately* after something. Parallel distinctions can be made between "before" (or "prior to") and "preceding." Consider the differences in temporal denotative value among the following sentences:

Ex. 1.48
1. The animals foraged only *after* they established their territory.
2. *After* [or *following*] 30 seconds the mixture turned blue.
3. Intraperitoneal injection was *followed* by activity testing.
4. *Following* [or *after*] intraperitoneal injection, we tested activity.

AMONG AND BETWEEN

Use "between" when referring to *two* things and "among" when referring to three or more, as in these two cases:

Ex. 1.49
1. There are key differences in clan structure *between* chimps and apes.
2. There are key differences in clan structure *among* chimps, apes, and baboons.

CAN AND MAY

"Can" connotes *ability to,* while "may" connotes *possibility* or potential. These two sentences show the difference:

Ex. 1.50
1. Neurons *can* grow longer, but our data imply that they also *may* reproduce.

2. Baboons *can* hunt for long periods, and they *may* forage over larger areas than previously observed.

COMPARE AND CONTRAST

We "compare" two or more things to demonstrate their similarities and "contrast" them to highlight their differences, as shown here:

Ex. 1.51
1. In our *comparison* of nesting patterns, we found a few minor *contrasts*.
2. Their roundness is *comparable* but their coloration *contrasts* sharply.

COMPLEMENTARY AND COMPLIMENTARY

A "complement" is something that makes up or completes a whole. (Two other technical senses refer to a group of proteins involved in antigen-antibody reactions and an angle related to another so that their total measures 90 degrees.) In contrast, a "compliment" is an expression of approval, praise, or civility. These senses are illustrated here:

Ex. 1.52
1. Genes have a specific sequence of *complementary* base pairs.
2. The animal consumed its full *complement* [total supply] of food.
3. We received a *complimentary* [free] supply of the medication.
4. We *complimented* the chimp to elicit the same behavior.

COMPOSE, CONSTITUTE, AND COMPRISE

"Compose" and "constitute" both mean *to make up,* but "comprise" means *to contain.* The whole comprises the parts, but the parts constitute or compose the whole.

Ex. 1.53
1. We *composed* the mixture slowly to avoid severe thermal effects.
2. The three inorganic *constituents* of the mixture are Na, K, and Ca.
3. The local bat population *comprises* [or *constitutes*] three species.
4. Cytosine, thymine, adenosine, and guanine are the bases that *compose* [or *make up,* but not *comprise*] the helical structure of DNA.

CONDUCT, DO, AND PERFORM

Of "perform," Robert Day notes aptly: "An unsuspecting person might think that scientists are monkeys or some other kind of circus animal. They are always *performing*. Some day, I hope that scientists will no longer *perform* experiments. It is much better to do them and be done with it."[37] Though "conduct," which has a range of connotations, is often used as a synonym, it seems pretentious. Indeed, "do" is best and simplest.

Ex. 1.54

1. We have done a study [versus *conducted an investigation*] to test that theory.
2. We have been doing [versus *performing*] experiments to find the cause of that effect.
3. One technician currently must do [versus *perform*] the work of two.

CONSTANT, CONTINUAL, AND CONTINUOUS

Something is "constant" if it occurs, behaves, or holds true steadily, unceasingly, or invariably. "Continuous" means one after another—extended or prolonged—without interruption or pause. "Continual" describes a discrete event that keeps repeating.

Ex. 1.55

1. For them to stay alive, water must *constantly* pass over their gills.
2. Our *constant* observation was that their courtship call is *continuous* for 5-second intervals and that they make it *continually* until they elicit a response from a prospective mate.

DIFFERENT FROM VERSUS DIFFERENT THAN

Since "from" is a preposition, *different from* is correct when followed by a prepositional phrase. In contrast, "than" is a conjunction to be followed by a clause. When unsure, opt for "different from."

Ex. 1.56

1. The nesting behavior of bluebirds *is different from* that of cardinals.
2. Sleep-deprived animals behave more aggressively *than* animals whose sleep patterns are not disrupted.

---

3. This is a *different* theory for explaining alcohol selection in hamsters *than* the one we proposed earlier.
4. For the squirrel monkeys in our study, the Pap Smear procedure was done *differently than* for humans.

FARTHER AND FURTHER

"Farther" connotes physical *distance,* while "further" connotes something *more.* Their contrasting senses are illustrated here:

Ex. 1.57
1. Prairie dogs range *farther* from their nests than do rabbits.
2. There is nothing *further* to be gained from such an approach.
3. *Furthermore,* once the animals became acclimated to their new habitat they protected it aggressively.

FEWER THAN AND LESS THAN

"Fewer" is used with something that is countable, while "less" is used with quantities and qualities (there are parallel uses of "number" versus "amount").

Ex. 1.58
1. The animals had *fewer* seizures with GABA than with a placebo.
2. The behavior of felines is *less* modifiable than that of canines.
3. Since it is raining *less* today, there is also *less* need for cover.

IMPLY AND INFER

"Imply" means to *suggest* or express something indirectly, as well as to entail. "Infer" means to *conclude,* to guess, or to have as a logical consequence. However, these senses can be conflated and the words used synonymously since an inference may also be taken to be suggestive or to have certain implicit (unspoken) meanings.

Ex. 1.59
1. The helical structure of DNA *implies* [or *suggests,* or *entails*] its copying mechanism.
2. One key *implication* of our results is that schools should be tested regularly for both biological and chemical environmental toxins.

> 3. From their extensive trials, Johnson and Green have *inferred* that the risk of side effects is much higher than expected.

INTERSPECIFIC AND INTRASPECIFIC

"Interspecific" means *between or among* species, while "intraspecific" means *within* one species, as this sentence illustrates:

> Ex. 1.60
> We must await more studies of this gene's *intra*specific frequency in mice before hypothesizing about how its frequency may vary *inter*specifically among mice, hamsters, and rats.

ITS AND IT'S

"Its" is the possessive form of *it,* and "it's" is the contracted form of *it is.* Contractions are seldom used in scientific prose or in other formal writing.

> Ex. 1.61
> The chameleon changes *its* colors readily, but *it's* unlikely to do so in unfamiliar surroundings.

A similar error can be made with "your" and "you're," again one being a possessive and the other a contraction of *you are.*

MANY AND MUCH

"Many" is used in reference to numbers and "much" is associated with quantity or degree (a similar confusion may occur with "fewer" and "less").

> Ex. 1.62
> 1. When we housed too *many* animals per cage, they exhibited *much* more aggression than when housed singly.
> 2. After *many* more trials, the treated group showed a *much* lower incidence of infection than the control group.

NORMAL, STANDARD, TYPICAL, AND USUAL

"Normal," points out Katz, "refers to a very specific distribution of numerical values—a smooth bell-shaped curve of an equation of the form y =

Kexp($-x^2/2$)."[38] If one means something that is *natural* or *habitual* or *ordinary* or *customary,* for instance, then that is what one should say. "Usual" refers to something—an event, a quality, or an entity—that occurs or is observed regularly or commonly. "Typical" refers to something—a trait or characteristic or quality—that is peculiar to, defines, or identifies some particular or discrete kind, group, part, category, action, behavior, phenomenon, or other entity. A "standard" is an agreed-upon or acknowledged measure of comparison, a *norm* or *criterion,* for quantitative or qualitative purposes and denotative value. These examples tell the four terms apart:

Ex. 1.63
1. Due to the lengthy rain season preceded and followed by brief periods of dryness, the week-to-week food supply is almost *normally distributed* [a bell curve] over the year.
2. We made the *usual* observation of nocturnal foraging but we could not yet characterize it as a *typical* behavior and hence as a *standard* to go by.

PRINCIPAL AND PRINCIPLE

"Principal" means a key or most important thing, while a "principle" is a basic truth or rule. The difference is demonstrated in this pair of sentences:

Ex. 1.64
1. In principle [*as a rule*] the liver is not affected, but there is a rare hepatic effect that constitutes a small risk of cirrhosis.
2. Our principal [*key,* or *primary*] finding is that the principle of fight-or-flight does not apply very well to the timid guinea pig.

THAT AND WHICH

The difference between "that" and "which" is in restrictivity: "that" introduces a *restrictive* element in your meaning, and "which," which usually follows a comma or a dash, introduces something *nonrestrictive*—that is, words that do not limit but instead *add* to the meaning. Note here the restrictive and non-punctuated use of "that":

Ex. 1.65
1. The *animals* that were *treated* recovered uneventfully.
2. We replaced the *fluid* that the animals *consumed.*

"That" restricts the meaning so that, in the first sentence, the author is referring only to *treated* animals, and in the second sentence only to *consumed* fluid. "Which" sets off information that "is not vital to the integrity of the sentence," the biologist Victoria McMillan notes, so omitting it will not substantially change the intended meaning. In the following two sentences, however, McMillan shows how the misplaced use of "which" results in ambiguity.

Ex. 1.66

1. The rats, which were fed a high-calorie diet, were all dead by the end of the month.
2. Plants, which grow along heavily traveled pathways, show many adaptations to trampling.[39]

In each sentence, the writer intended a restricted reference—to the *subset* of rats "fed a high-calorie diet," and to plants growing "along heavily traveled pathways." Instead, both sentences appear to make generalized references to all rats or plants. To convey the intended meaning, the "which" in both sentences should be changed to "that" and the commas should be deleted. Finally, in the following two sentences, the first version refers specifically to a subset of DBA mice that "metabolize acetaldehyde slowly" and the second refers generally to *all* DBA mice as slow metabolizers.

Ex. 1.67

1. Mice of the DBA strain *that* metabolize acetaldehyde slowly drink significantly less ethanol than mice of the C57 strain.
2. Mice of the DBA strain, *which* metabolize acetaldehyde slowly, drink significantly less ethanol than mice of the C57 strain.

VARIOUS AND VARYING

"Various" refers to different kinds, while "varying" means changing. This sentence distinguishes their uses:

Ex. 1.68

We *varied* our position daily and inferred that the *various* species of birds were in fierce competition due to the *varying* food supply over the year that has constituted an unusual seasonal *variation* during the decade.

One must be careful not to use the noun form in redundant constructions like: "We saw a variety of different bird species."

Besides the sampling provided here, there are various other pairs or clusters of misused words. One example is the erroneous use of "since" (which has a temporal connotation) in place of "because" (which is causative). "*Since* we could not quantify the intermediate metabolite accurately, we instead determined . . ." leaves the reader with a potential ambiguity; the sentence is more precise if it begins with "Because."

PUNCTUATION

It is fitting here to consider the role in expository writing of punctuation, not as an afterthought but rather because it is an aspect of communication that, first, stands *outside* of words and, second, affects all aspects of writing—clarity, simplicity, preciseness. Recall for instance the cues that commas provide when used with "which" to introduce a nonrestrictive clause (Exs. 1.66 and 1.67), or their usefulness in references to series of entities. The various kinds and lengths of pauses possible in written communication—from the *short* pause of a comma or colon to the *intermediate* pause of a semicolon and the *full* pause of a period—permit clarifying and simplifying nuances of scientific denotation. The physician and researcher Lewis Thomas offers a tongue-in-cheek illustration of the function of punctuation in this lengthy sentence that develops in parenthetical layers.

Ex. 1.69

There are no precise rules about punctuation (Fowler lays out some general advice (as best he can under the complex circumstances of English prose (he points outs, for example, that we possess only four stops (the comma, the colon, the semicolon, and the period (the question mark and exclamation point are not, strictly speaking, stops; they are indications of tone (oddly enough, the Greeks employed the semicolon for their question mark (it produces a strange sensation to read a Greek sentence which is a straightforward question: Why weepest thou; (instead of Why weepest thou? (and, of course, there are parentheses (which are surely a kind of punctuation making this whole matter much more complicated by having to count up the left-handed parentheses in order to be sure of closing with the right number (but if the parentheses were left out, with nothing to work with but the stops, we

---

> would have considerably more flexibility in the deploying of layers of mean-
> ing than if we tried to separate all the clauses by physical barriers (and in the
> latter case, while we might have more precision and exactitude for our
> meaning, we would lose the essential flavor of language, which is its won-
> derful ambiguity)))))))))))).[40]

Although this humorous example does focus on the uses of punctuation, the
declarative nature of scientific writing favors sentences that are short, direct,
and simply constructed, with punctuation used conservatively and sparingly.
Periods and commas are basic necessities, but because simpler and unconvo-
luted sentences require fewer clauses, even commas should be used with restraint. The other marks illustrated by Thomas (the parenthesis, semicolon,
and question mark) as well as those not illustrated—the exclamation point and
the dash—are used less often in scientific prose. In that sense, the relative frequency of the various punctuation marks in Thomas's paragraph, with the
salient exception of parentheses, is closely representative of their usage frequency among researchers. Along with typographical errors, punctuation mis-
cues are among the most serious threats to a scientific document's denotative
clarity and precision. They are also the most readily avoidable problems that
scrutiny of a draft can detect. As with any other aspect of effective writing,
punctuating is an art that writers can improve at with experience.

SCIENTIFIC ENGLISH AS A DYNAMIC INSTRUMENT

This chapter began by focusing on how scientific uses of language have historical and philosophical roots that inform the way researchers write today.
These roots, together with the academic study of the writing process itself, reveal scientific English as dynamic, collaborative, and highly formalized in its
practice. Its vibrancy and technical rigor derive from several factors. First, its
unique features and restricted application make it an essential instrument of
objective inquiry that must continually be honed and inspected so that it remains reliable and effective. Honing this instrument is a fluid and ongoing
professional process. Second, the instrument itself as well as what it produces—researchers' writing—is dynamic because it is human, social, and
continually evolving as a shared tool among its users. Just as scientific research is a social enterprise, the language that sustains it must constantly and
recursively go through a process of communal scrutiny, reevaluation, editing,

and change. There is a constant interplay between the *individual* human dimension of scientists-as-writers and the limited range of linguistic freedom allowed by the research *community* to which they belong and to whose members they must ultimately answer. The use of scientific English therefore constitutes a dynamic process that is cognitive, social, and cultural, the latter in its senses of both professional and global community.

One defining quality of the critical social and human energy of scientific English is that it abounds with the inventiveness and ingenuity of neologisms. Wilkinson underscores the constant invention of new language in the sciences: "New words are such a regular part of scientific research and so generally accepted that coining them is not recognized as a scientific achievement. Furthermore, such coinages have become an increasingly common activity among scientists, because scientific research is a more widespread activity and because of the increasing complexity of modern research." While scientific neologizing has picked up its pace in our time with the accelerated growth and sophistication of experimental research, out of practical necessity this inventive spirit has always been an integral part of scientific work and thought. The nineteenth-century chemist Michael Faraday exemplified this spirit in his coinage of numerous scientific terms, many of which are still used today, such as the ones he mentioned in this excerpt from a letter to a friend in 1834: "I wanted some new names to express my facts in electrical science without involving more theory than I could help, and applied to a friend Dr Nicholl, who has given me some that I intend to adopt. For instance, a body decomposable by the passage of the electric current, I call an *'electrolyte,'* and instead of saying that water is *electro chemically decomposed* I say it is *'electrolyzed.'*" Besides "electrolyte," among the other words Faraday gave us are electrode, anode, cathode, anion, cation, ion, diamagnetism, and paramagnetism.[41]

Scientific language evolves collaboratively and socially in the community of researchers. Sometimes a new word or term arises from a new scientific perspective that the old word could not denote. The following sentences illustrate a shift in usage among alcohol researchers since the 1970s.

Ex. 1.70
1. Drugs that inhibit ethanol *self-selection* in animals may also reduce intake of other *preferred* fluids such as saccharin solutions.
2. Mice of the high-ethanol-*selecting* C57/BL/2j strain consume signifi-

---

> cantly larger amounts of 10% solution of 1,2-propanediol and 1-propanol than the low-ethanol-*selecting* DBA/2j strain.
>
> 3. Mice from the high-ethanol-*preferring* C57BL strain and the low-ethanol-*preferring* DBA strain were tested for their preference for butanediols.
> 4. We have previously observed that a pharmacological dose of Ach (50–150 mg/kg) administered intraperitoneally (i.p.) produced a dose-dependent flavor aversion in low-ethanol-*drinker* (UchA) rats, whereas high-ethanol-*drinker* (UchB) rats appeared to be insensitive to Ach.[42]

The terminology used to denote the animals' ethanol consumption behavior has changed progressively from "ethanol-preferring" to "ethanol-selecting" to the currently prevalent "ethanol-drinker." The first sentence actually mixes the first two terms. Over time, however, the prevalence of the most neutral term, "drinker" (sentence 4), represents a move away from earlier anthropomorphic expressions of "preference" or "liking," which are associated with human taste or pleasure. It is understood among researchers that as any field develops, its terminology will evolve along with new and changing perspectives and denotative needs.

The human vitality of scientific English also resides, as Luria pointed out, in the natural individuality of style through which writers connect effectively with their readers. In the chapter "Expressing Ideas and Reducing Bias in Language," the APA *Publication Manual* tells researchers: "Thoughtful concern for the language can yield clear and orderly writing that sharpens and strengthens your *personal style* and allows for *individuality of expression* and purpose." Whatever individual form a researcher's style may take, that chapter in the APA manual asserts that, in addition to being clear and orderly, effective research writing must also be precise, logical, smooth, and agreeable. The reserved degree of individual latitude in human expression that scientific English does permit is more than sufficient to demonstrate that a researcher's writing process is not mechanical and lifeless but rather vigorously alive and evolving.

Finally, scientific English is fundamentally dynamic because the writing that researchers do bustles with action. It describes both what they do procedurally and the outcome of that methodology—that is, what they see *happening* as a result of their experiment. There is a necessary and perpetual presence of activity and vitality in scientific writing that is typified in research papers.

Action permeates scientific English. This is so from the get-go in scientists' primal application of language to record actions and reactions—something they begin learning in undergraduate laboratory classes—in the course of their research.

Researchers are both writers and readers. The chemists Hans Ebel, Claus Bliefert, and William Russey note: "It goes without saying that scientists need to be skillful *readers*. Extensive reading is the principal key to expanding one's knowledge and keeping up with developments in a discipline. The often ignored corollary to this assertion, however, is that scientists are also obliged to be skillful *writers*. Only the researcher who is competent in the art of *written* communication can play an active and effective role in *contributing* to science." From the perspective of readability, moreover, scientists should always write with a reader-centered mentality; even in the act of writing they must be mindful of the act of reading. "In order to understand how to improve writing," Gopen and Swan assert, "we would do well to understand how readers go about reading." It is this key realization that writing and reading are two sides of the same linguistic coin that serves as the best guide for using scientific English effectively. The agronomist Martha Davis, who has instructed fledgling scientists on scientific writing's challenges, offers this metaphor: "Writing or speaking about scientific research is no more difficult than other things you do. It is rather like building a house. If you have the materials you need and the know-how to put them together, it's just a matter of hard work." As you wield that mighty tool of scientific English in the course of constructing your own house, remember to continually evaluate the practical functionality of that house for its prospective occupants—your dear readers.[43]

2

LABORATORY NOTES

We are all liable to error, but we love the truth, and speak only what at the time we think to be the truth; and ought not take offence when proved to be in error, since the error is not intentional, but be a little humbled, and so turn the correction of the error to good account.
—Michael Faraday, letter to C. Matteuchi, 1855

PURPOSE OF LABORATORY NOTES

One of the very first and most fundamental applications of the language of science is in keeping notes on laboratory research. Such notes are a basic part of sound experimental work. Whether in early science curricula, in college science programs, or in professional research settings, those who are engaged in experimental inquiry should understand and apply the stringent principles of maintaining a written record of their actions and thoughts during the course of that process. Researchers use laboratory notes for various purposes. The most important of these is to record their experimental design, methods, observations, and results. To this recorded information one can then add analytical notes that discuss, evaluate, and interpret the observations and results to reach whatever conclusions are permissible. Supporting materials, such as printouts from instruments or photographs, naturally must also be safely pre-

56

served. If laboratory notes are written with sufficient clarity and detail—along with the requisite accuracy and precision—they can be used in the future for verification and replication. In addition, well-kept notes facilitate their use for subsequent purposes, such as writing laboratory reports or professional papers, applying for patents, or planning further research. Hans Ebel and his colleagues assert that a properly kept laboratory notebook is "arguably a scientist's most important tool. Working in a laboratory—even pondering a complex set of ideas in one's office—without having a notebook open nearby should be unthinkable, unnatural."[1] This chapter focuses on the principles and process of writing and managing a laboratory notebook, with due attention to how the notes are used to write laboratory reports.

NOTEBOOKS IN THE WORKPLACE AND EDUCATIONAL SETTINGS

Keeping a laboratory notebook in any research setting, whether in industry, government, or academe, is an essential workplace or educational responsibility. In the workplace, the information recorded in notebooks is helpful in various ways. It permits colleagues to assess one another's work in collaborative situations, facilitates supervisory reviews of an employee's research progress, and provides critical documentation for patents. Industrial laboratories adhere to strict protocols for thorough review, including witness signatures, of notes and data by colleagues or supervisors. This is helpful not only in patent applications but also as an effective protective measure against fraud in scientific research. In higher education, students keep a notebook because it is required either by a course or for graduate research leading to a thesis or dissertation.

Undergraduate students majoring in the experimental sciences are expected to learn forms of writing that are based directly on their own research experiences, namely, laboratory notes, laboratory reports, and research reports. In its current guidelines for Undergraduate Professional Education in Chemistry, for instance, the American Chemical Society (ACS) notes the critical importance of writing based on laboratory experience. In the section "Laboratory Work in Chemistry," the ACS guidelines state that, in gaining hands-on experience with chemistry, students should also acquire "the self-confidence and competence to keep legible and complete experimental records" and to "communicate effectively through oral and written reports." The guidelines emphasize the importance of laboratory research experiences that culminate in a research report: "A well-written, comprehensive, and well-documented research report must be prepared, regardless of the degree of success of a student's project. The faculty supervisor should constructively criticize the report during the draft stage. Oral, poster, and computer presentations do not meet the requirement of a comprehensive written report. Student co-authorship on a journal article, while highly desirable, is not a substitute for a comprehensive report written by the student." Although college reports typically are based solely on research in the scientific literature (described in Chapter 4), the form referred to here synthesizes the results of the student's own *laboratory* research with *bibliographic* research in the journal literature and is focused on the experimental project's topic and hypothesis. The value of other forms of scientific communication notwithstanding, they do not achieve the highest level of critical thinking in scientific writing, which only reality-based and applied experience can teach—simply put, reading and writing that is focused on one's own laboratory project. This full spectrum of research, in its senses of both laboratory and bibliographic, fully teaches how science works.[2]

In undergraduate courses, science students are likely to be required to keep a laboratory notebook, especially in their upper-level classes. It is common for laboratory instructors to provide supplementary materials having detailed guidelines for writing both laboratory notes and laboratory reports. For undergraduates, acquiring the habit of proper note taking will not only lay a smoother path toward laboratory reports but in the longer run will also provide a valuable competence applicable to any job. For those pursuing a graduate degree, effective notes will facilitate the writing of a thesis or a dissertation, or an article, and even offer protection in cases of invention (such as new techniques or hardware) that lead to a patent application. The importance of acquiring the discipline and sound habits associated with laboratory note-keeping—a challenging and sometimes even a resisted activity—cannot be overemphasized.

LEGAL AND ETHICAL RESPONSIBILITY IN LABORATORY NOTES

Strict and rigorous adherence to the routine of maintaining proper laboratory notes also carries legal and ethical implications. For instance, consider a situation in which competing laboratories or researchers develop an innovation a few months apart and all parties concerned apply for a US patent. This is a legal circumstance that can be resolved through corroborating evidence contained in properly and meticulously written laboratory notes. Legally, it may be difficult to show who had an *idea* first—unless the researcher's notes show some form of "reduction to practice." In other words, the laboratory notes must show evidence that something was *done* in association with the idea, such as the development of a working model, the isolation of a compound, or the application of a new technique. Sound note-taking practices will ensure that one can support a claim to having legal rights over an innovation. To be valued legally as corroborating evidence, notes must also be accepted as being unaltered and credible.

Besides the legal considerations, there is the matter of basic ethics. On rare occasions, professional pressures can lead a researcher to succumb to the temptation of wrongdoing. This may lead to some form of unethical notekeeping. The physicist-writer C. P. Snow's novel *The Search,* intended as a realistic glimpse into the demanding life of a research scientist, offers a compelling scenario to illustrate this point. At a critical moment in the story, young chemist Arthur Miles sees in one of his x-ray crystallographs an anomaly fatal to his hypothetical model of "the structure of the organic group." In this painful moment of truth, given the importance of confirming his major finding, Miles must confront his temptation to deny the anomalous finding. What if he had not taken that particular photograph? Without it, the evidence supporting his hypothesis was overwhelming. Should he destroy it? Even should the error finally be discovered, long after his paper is published and his reputation made, he could just claim that it was an honest mistake. Courageously, and in keeping the highest standards of scientific ethics, Miles makes the following notebook entry: "Mar. 30: Photograph 3 alone has secondary dots, concentric with major dots. This removes all possibility of the hypothesis of structure B. The interpretation from Mar. 4–30 must accordingly be disregarded."[3] His entry is, of course, a victory for science, a stand against the fraud that sometimes creeps into laboratory research, like the failure to note inconvenient facts.

Or, consider an actual instance of suspected dishonesty in a 1986 article in the journal *Cell,* a case that hinged on 17 pages of laboratory notes from the hundreds of pages generated by the project. An accusation by a postdoctoral fellow of faked experiments and altered notes by a team member mushroomed until it became the subject of prominent media attention, review by various ethics boards, and congressional hearings in Washington, DC. The affair culminated in 1991 when some of the authors (including the principal investigator) sent a letter to the editor of *Cell* to retract the paper. Attempts by others to duplicate the results were unsuccessful.[4] Whatever the motivations may be for the relatively few cases of scientific dishonesty—ambition, glory, the rewards of recognition—sooner or later the experimental record left in the laboratory notes, or lack thereof, is likely to reveal its own story. Without impeccable ethics in the laboratory and in notekeeping, the progress of science itself is in jeopardy. In addition to the critical standards associated with ethical and legal concerns, well-maintained laboratory notes must be complete, clear, accurate, precise, and authenticated. In the process, a researcher must give due consideration to their permanence and careful organization.

PERMANENCE OF NOTEBOOKS AND NOTES

To begin with, notes must be kept in a manner that allows them to be accessible and usable permanently. In any professional research setting, it is not uncommon for the need to arise to consult notes for various reasons even more than a decade later, and in patent disputes more than twice as long after they were taken. The chemist Howard Kanare has suggested that "we should be concerned about maintaining original research notes for at least 25–30 years; the paper should be in such good condition that it can be handled and studied without fear of damage to the physical record. At the same time, the writing must be in such good condition that it can be read and understood without ambiguity."[5] To allow for such permanence, certain basic tools and practices must be used. The specific items of concern are the type of notebook, the quality of paper, and the most suitable writing implement. Laboratory notes are typically kept in a bound hardcover notebook, rather than on such alternatives as separate pieces of paper, a looseleaf binder, a spiral notebook, or even a student composition book. A case-bound notebook is best because its pages are sewn and glued together, and it can be laid flat without concern for wear and tear. With a well-bound notebook, the pages cannot be separated or their order cast into doubt or, worse yet, simply lost. Bound notebooks designed especially for laboratory purposes are readily available in student bookstores and through online outlets. A specialty outlet will provide specific information about the qualities just mentioned as well as those of the paper.

While the binding keeps everything firmly together and gives the notebook durability, Kanare notes that the paper must also have characteristics that give it longevity. Key features associated with the longevity of writing paper include the purity of the wood pulp, amount of lignin, acidity level, and rag content. The most suitable paper for notekeeping is composed of 100 percent chemically purified wood pulp. In addition, the best paper contains no lignin or ground wood and no alum-rosin sizing agent, with a minimum pH level of 5.5. In addition, paper with about 3 percent calcium carbonate as an alkaline reserve will last longer. If the notebook supplier is unable to confirm the paper's composition, spot tests can be done using a paper testing kit.[6]

Besides a bound notebook and paper of durable quality, appropriate writing implements must be used. Kanare underscores that the best writing tool for ensuring permanence of notes is a ballpoint pen with a fine tip and black ink. Color inks, especially red, are more sensitive to light exposure. Avoid using porous felt-tip, plastic ball roller-tip, and fountain pens. The ink should be fast-drying, stable in the long term and against light, resistant to chemical degradation, nonreactive with paper, and easily microfilmed or photocopied. Notes taken with a lead pencil have several problems: they can smudge, photograph poorly, and erase easily (leaving their integrity open to question).

Along with the notes, a researcher may attach such supporting materials as photographs, sheets from analytical instruments, and photocopied materials including letters, memoranda, proposals, and journal articles. For such attachments, Kanare advises archival quality mending tape (not the common office supply variety) or high-quality, acid-free white glue should be used, to prevent an adverse reaction with paper or ink. Attached photocopies must be on high-quality paper, and tape or adhesives should not make direct contact with the printed images. The tape or glue should be used sparingly and the attached materials pressed so they flatten sufficiently to prevent crimpling later. As much of a bother as it may seem at the time, taking these simple additional steps to ensure permanence in recordkeeping will repay itself many times over whenever the need may arise to consult or recheck the notes.

NOTEBOOK ORGANIZATION AND ENTRIES

The simple but highly effective conventions associated with laboratory notekeeping are applicable across the educational and occupational spectrum of settings in which scientific experimentation is taught or practiced. Whatever the setting of the laboratory research activity, the same key criterion holds for assessing the effectiveness and value of a set of laboratory notes: Does the notebook contain a full, reliable, and accessible record of what its author did in the laboratory? To ensure the reliability and usability of the scientific narrative that the notes tell, standard practices are followed for writing and organizing a notebook. Though there are setting-specific expectations, such as those associated with the notebook as a corporate and legal document, the basic tradition is the same everywhere. Notebooks contain some front matter, such as a table of contents, followed by the notes themselves in dated, sequential entries. The pages of the notebook should be numbered at the upper right, have a heading with space for your name, the date, the subject, and, especially for industrial settings, for witness signatures. Having your notes reviewed and signed by witnesses is important not only legally but also as a sound practice of scientific double-checking, which fosters maximally objective and error-controlled research that is demonstrable in the written record.

FRONT MATTER AND GROUND RULES

The items in a laboratory notebook's front matter, as well as some of the ground rules for preparing entries, vary somewhat across settings. Here is a list of items that, in one setting or another, are commonly included in the front matter:

- Cover title
- Sign-out page
- Instructions page
- Table of contents
- Preface
- List of abbreviations

The first item of front matter usually encountered by a notebook's readers is its external identification or *title,* which should be written either on the front cover or on the spine with easily readable and durable ink. Titles may consist of a project's subject, such as "Ethanol Metabolism," or simply a numerical designation. One effective option is to combine your initials with a *volume* number in roman numerals (to distinguish it from the arabic page numbers), for example, RCG-II. This approach also provides a simple system for *cross-referencing* among notebooks and related experiments. If *experiment numbers* are used (corresponding to the starting page of notes for an experiment, for instance), a particular experiment might be referred to as RCG-II-36. Using this system, one could even refer to individual substances or fractions. Thus, chromatography fraction 5 from experiment RCG-III-96 might be identified as

RCG-III-96-5, so long as all coded items are carefully identified in the notebook. The convenience of such identification codes may be extended to the labeling of vials or spectra. A system like this combines brevity with specificity, all rooted in the notebook's title.

Beyond a title, two items of front matter that are standard in corporate and government notebooks are a *sign-out page* and an *instructions page*. As the term implies, the sign-out page provides spaces for the names and signatures of the notebook's issuer and recipient, the date issued, and the dates on which the notebook was completed and submitted, again with the appropriate signatures. As if these workplace signatures did not make a notebook official enough, for legal purposes it is also necessary to have witness signatures *internally* with the experimental entries when any important experimental outcomes may lead to a patent application. Following the sign-out page, the notebook provides an instructions page for its users, stating that—as the employer's property—the notebook's entries are to be prepared with strict adherence to certain occupational ground rules.

The next standard component of a notebook's front matter is a *table of contents,* for which several blank pages are typically reserved so that a list of the contents can be added once the notebook has been filled. A table of contents allows for quick location of specific areas of interest in the notebook, so long as it is kept accurate and current. The items listed in the contents vary according to setting and purpose, but the basic log consists of an entry's date, subject, and page numbers, as in the following example of a graduate student's notebook.

Ex. 2.1

TABLE OF CONTENTS   Notebook No.: *RCG-1*   *MS Thesis Research*

Date	Subject	Page no.
Jan. 17, 1974	*Preface (CNS sensitivity to alcohol in mice)*	*4–5*
Jan. 18, 1974	*Activity tests (after butanediol injection)*	*6–8*
Jan. 19, 1974	*Alcohol metabolism (trial runs with Cary 14 Spec)*	*9–12*

The page also includes the notebook number and its identification as thesis research. Additional information can be logged for setting-specific purposes, so that tables may also have columns for such items as project numbers, product codes, or client names. Other than dates and page numbers, there is much flexibility for adapting a table of contents to the specific purposes and needs of any situation.[7]

Immediately following the table of contents—and the first item one might log on it—is the oft-neglected *preface*. A few prefatory remarks about you and your purpose as the notebook's keeper will orient any prospective reader. The preface's contextual information may include your name, the location of the work and your capacity as the notebook's writer (course or job title, for instance), and the project's purposes, goals, and relation to any prior work.

As a final item of front matter, one may include a *list of abbreviations* that are used in the notebook. This list is essentially a glossary for shorthand or coded terms that refer to such things as supply companies and their products, laboratory equipment, chemical preparations, and experimental samples. A researcher who refers to two different genetic strains of laboratory mice obtained from Jackson Laboratories in Bar Harbor, Maine, for instance, might use such abbreviations as C57BL/6J (high-alcohol-selecting strain) and DBA/2J (low-alcohol-selecting strain), along with an abbreviated reference to the vendor, JLBH. The list also may include such items as standard reagents, statistical procedures, and special equipment. Like the table of contents, a list of abbreviations is an item of convenience to be adapted to one's situational needs.

A notebook's front matter is a valuable aid for its users that is well worth the extra few minutes of attention upon starting a new notebook. The same extra care is needed in following the conventional expectations for writing the notebook's entries. Here are a few simple and commonly practiced ground rules:

- Start new research on a new page, and write on only one side of the page.
- Date and initial each page.
- Do not skip or remove pages, but draw crossed lines to void unused space.
- Write entries during the course of the experiment, not by relying on memory later.
- Delete or correct notes when necessary by crossing out, explaining, and initialing.
- Add new thoughts to preceding records only by making cross-referenced entries.
- For collaborative projects, with multiple note takers, agree on a common note-taking system that all will use.

- Request signatures of witnesses for professional and legal purposes.
- Write legibly and plainly to permit reader accessibility.

Following these conventions habitually will permit you to keep your notes organized, readable, and credible. Beyond the practical ground rules, the experimental entries themselves must be written using conventional elements of content and organization.

CONTENT AND STRUCTURE OF NOTEBOOK ENTRIES

What are the conventional features of an experimental entry in the notebook? Standard practice is to organize the entries using the following sequence of components:

- Introduction and background
- Methods and materials
- Observations and results
- Discussion and conclusions

Some of the details of content or format within these four major components of a notebook entry will also need to be adapted for experimental work that is done in the field, such as with plants, insects, or animals. Field notes require detailed descriptions and diagrams of experimental locations, for instance, with precise dimensions, qualities, and conditions. Whether an experiment is done internally or in a field laboratory, each of the four components plays a vital role in the completeness and integrity of notes and demands close attention to their required details.

Introduction and Background

Beginning on a new page, and following a dated heading that identifies the new project, it is good practice to explain the scientific problem being investigated and how it is addressed by the planned experimental work. The introduction should start with a clear, simple, and specific statement of your purpose, such as in this example for an alcohol experiment:

Ex. 2.2
The purpose of this experiment is to measure the drinking rates and neural effects of the four isomers of butanediol in alcohol high-selecting (C57BL/6J) and low-selecting (DBA/2J) strains of laboratory mice.

The rest of the introduction then elaborates as needed, depending on the project's scale. Is it a single or simple experiment to last hours or a more complex project needing a series of experiments over weeks? How does the work fit into its wider context of related studies? What were the previous results? What citations or cross-references to your own work are useful? Why did you choose this project over other options? How would the anticipated outcome be significant or beneficial? Experiments take up precious resources—human, temporal, spatial, and economic—so they must be chosen wisely to answer worthy questions. The experimental purpose articulated in Ex. 2.2 could be developed as follows:

Ex. 2.3

Genetic differences in ethanol selection by laboratory mice were discovered by McClearn and Rodgers (1959). The DBA strain is low-selecting and the C57 strain is high-selecting. Possible causal factors discussed by McClearn (1972) include differential neural sensitivity to ethanol and rate of ethanol metabolism in the liver. Interstrain differences in consumption, neural effects, and metabolism have been found with ethanol and with two other alcohols, 1-propanol and 1,2-propanediol (Strange et al., 1976). However, no such studies exist with the butanediols to be used in this experiment. A finding of significant differences in consumption rates and neural sensitivity using another alcohol, butanediol, would support the hypothesis that inherited neural differences result in potential differences in drinking behavior. If so, then mice of the high-selecting C57 strain would be expected to be less sensitive than mice of the low-selecting DBA strain to the biphasic effects (i.e., stimulation followed by narcosis) of the four isomeric butanediols to be tested.

The statement in Ex. 2.3 is concise but informative, cites related work, and provides a theoretical framework (inherited neural sensitivity) for the planned research. The level of elaboration in introductory notes varies and may include details on important chemical properties or reaction equations, and sketches of innovative equipment or techniques. For Exs. 2.2 and 2.3, details could include structural formulas for butanediols, metabolic pathway of butanediol, or a sketch of equipment used to test neural sensitivity to alcohol. When consulting the notes later, the notebook's writer and other readers will be grateful for the contextual details that introduce the experimental undertaking. Only after these preliminary notes are completed does it make sense to focus on describing the experimental plan itself.

Methods and Materials

The methods and materials section of the notes provides all the details of the experimental design, materials, and procedures. Although it is useful to outline the experimental procedure here, this outline cannot take the place of a narrative description later of what actually was done and observed. This description of the experimental plan should be preceded by a list of the materials and other relevant items, such as the following:

- commercial and noncommercial materials and resources, including chemicals and animals, with supplier, lot number, grade, packaging, age, and expiration date;
- chemical names, formulas, and properties (e.g., molecular weights, melting point and boiling points, solubility, specific gravity, toxicity, viscosity, color);
- important instrumental parameters, calibrations, and measurement conditions;
- laboratory conditions, including such external factors as temperature, humidity, lighting, air quality, and pressure, noting any fluctuations during the experiment;
- equipment, with sketches of any unfamiliar, modified, or innovative features, and manufacturers, models, or catalog numbers.

Here is a sampling of listed items for the alcohol project introduced in Exs. 2.2 and 2.3 that includes the animals, neural testing equipment, and alcohol solutions:

Ex. 2.4

1. ANIMALS: Mice from Bar Harbor Labs, Maine, 300 total: 150 of high-drinker strain (C57BL/6J; black fur) and 150 of low-drinker strain (DBA/2J; gray fur); 10–12 weeks old; male; kept in standard metal cages; fed standard diet; lab light cycle 8 a.m.–5 p.m., 68° F.
2. ALCOHOL SOLUTIONS: 10% (v/v) in distilled water of four butanediols (1,2-, 1,3-, 2,3-, 1,4-butanediol). From Dow Chemical, Midland, MI; used without testing for purity.
3. ALCOHOL TESTING TUBES: Kimax centrifuge tubes, 15 ml graduated in 0.1 increments, stainless steel spout with 2 ml orifice.
4. NEURAL TESTING EQUIPMENT: Jaw-jerk response apparatus (Schneider, 1973) and open-field activity apparatus (Hillman, 1975).

Note that when a special apparatus or technique has already been described elsewhere—in prior notes or publication, as in the fourth item (neural testing equipment) above—one may make cross-references or attach cited articles. When beginning a particular experiment within a larger project, it is helpful to make a more detailed list of the required materials, like this one for determining a mouse's rate of alcohol (1,3-butanediol) metabolism:

Ex. 2.5

SUPPLIES:
pipettes
ice tray, ice bowl (petri)
large tubes for ice tray (kept cold)
centrifuge tubes (plastic), 50 ml
dissection tray (wax)
dissection tools: pins, scissors
graduated cylinders
beakers
test tubes (5 ml)
plastic mouse cage
C57 and DBA mice (12 of each strain, T = 24)
filter paper for liver (7.0 cm), Whatman
magnetic stirrer
wax paper for cuvettes (mixing)

EQUIPMENT:
Sorvall Superspeed RC2-B Automatic Refrigerated Centrifuge
Cary Model 14 recording spectrophotometer (0.0–0.1 expanded scale)
Cary-14 cuvettes, matched
Spec 20 spectrophotometer
Potter-Elevhjem homogenizer
pH meter (329)
Open-field activity box
Jaw-jerk response apparatus

SOLUTIONS:
NAD—10 mg/ml
Sodium pyrophosphate buffer—pH 9.6
1,3-butanediol—3.9%
Liver homogenate—in 9 vol. sucrose sol.

0.25 M sucrose sol.—in dist. H_2O
Bovine serum albumin—1 mg/ml
Biuret Reagent

There is more detail on such a list than would be needed for a formal write-up of experimental work in a laboratory report.

Once all the materials are listed, the experimental plan must be described. This may consist of an enumerated, step-by-step, and concise listing of what is to be done and measured. When the experimental narrative is written later, to tell what actually *was* done, cross-references to these enumerated steps in the procedure are facilitated. That is, the researcher later can check the completed procedures against the steps originally described in the experimental plan, and can include parenthetical cross-references to that plan. This is especially helpful if any step in the original procedure is modified, such as with reagents, timing of measurements, instrumental settings, or experimental conditions. The level of detail typically needed in this list of procedural steps is modest, because the full procedural details will be written out later as they are being done. Moreover, detailed experimental protocols are commonly provided— for example, in the workplace for standard procedures or in educational settings by laboratory instructors—and may be attached to the notebook. In our alcohol project example, the experimental aim expressed in Ex. 2.3 to measure the consumption by mice of butanediols could be written out as follows.

Ex. 2.6
1. House each animal individually and allow 4-day acclimation period.
2. Test 15 mice from each strain (high- and low-selecting) for 10 days with 10% solutions of the four butanediols as a choice with distilled water (120 mice total).
3. Measure amount of fluid consumed from each tube every 24 hrs, at 10 a.m. Switch tubes (H_2O and alcohol) daily to avoid position effects.
4. Determine drinking (selection) index for each animal by dividing the amount of alcohol consumed by the total amount of fluid consumed (account for spillage error). Derive a mean index for each group by averaging the 10-day period.

Later, as each step of the plan is completed, the page numbers for those experimental notes can be added here. Similar procedural plans will need to be written for the other experimental components of the project. The level of detail a plan needs varies with the given situation and with the extent of reliance on cross-referencing, such as to prior notebook records, standard protocol documents, student worksheets in a laboratory course, or published sources. Greater detail may be necessary when an experimental design is complex or innovative. Once a plan is laid out and is being followed, the core of note writing has been reached, and the moment-to-moment activity both procedurally and with regard to experimental outcomes must be recorded.

Observations and Results

Laboratory time and space should be made to ensure that a habit of note-keeping is an integral part of the experimental proceedings. Most of the results section can be in the form of a narrative in the first person that describes what you did and observed as you tell the experimental story. Record your observations on the spot and completely, saving interpretation for the concluding section. Writing notes as you go along will spare you the cost of a distracted mind that is cluttered with facts better unloaded onto the pages of a notebook, as well as help you avoid forgetting crucial details when trying to recall them later. As you follow your experimental plan, you will record the outcomes to be evaluated later for their support or refutation of your hypothesis. In our alcohol case, the data will be interpreted to either support or refute a hypothetical link in laboratory mice between inherited neural sensitivity and drinking behavior, with high-drinkers being less sensitive than low-drinkers to the neural (stimulatory and depressive) effects of alcohol. Here is a sample data sheet in tabular form developed to record the rates of alcohol (1,3-butanediol) metabolism in the liver of a laboratory mouse.

Ex. 2.7 Buffer (ml)	NAD	Enzyme (dil)	Alcohol (BD)	$\Delta OD/$ min	SA ($\mu M/$ min/mg)
2.7	0.2	0.1 (1/100)	0.1	0.040	0.74
2.8	0.2	0.1 (1/50)	0.2	0.060	1.10
2.6	0.2	0.1 (1/20)	0.2	0.044	0.81
2.5	0.2	0.1 (1/1)	0.056	0.112	2.06

Specific activity (SA) for 1,3-butanediol (BD) recorded for C57 (high-selecting) mouse, wt. 24.0g, liver 1.3g, homog. T = 3 min, 12 wks age

The data contained in the caption, such as the weight of the animal (24 grams) and of its liver (1.3 grams), will have been recorded in the notes as the procedure was conducted but also can be included with the tabulated results for convenient reference. When describing the experimental procedure, one should also include a list of terms and symbols, such as "$\Delta OD/min$," a rate designation standing for "change in optical density per minute." One can also add a key for such terms to accompany the data. The value of a record of what is done and seen naturally will be a direct function of how objective, complete, clear, and scrupulously honest it is. Recording selectively may lead to the sort of temptation faced by young Arthur Miles in Snow's fictional scenario.

How much detail does an experimental narrative need? The level of detail required does vary, but it is better to be habitually thorough and comprehensive than to risk omitting any potentially important details. Note carefully in your narrative any unexpected observations or deviations from your experimental plan or from routine procedures, however slight. Seemingly trivial observations or modifications can turn out to be among the most important details of an experiment. Moreover, a detail that seems hardly noteworthy to you will be deemed otherwise by a reader who needs it to duplicate your results. It is better to err on the side of excess than to risk missing a key fact. As Ebel and colleagues put it, "The recollection of a peculiar color change on extracting an ethereal solution with aqueous alkali, finding a notation of a spur-of-the-moment decision to use NA_2SO_4 as a drying agent (because the $MgSO_4$ bottle was empty), being able to glance once more at the spectrum of a supposedly useless distillation residue—any of these might provide the key to a crucial insight."[8] Depending on a particular project's requirements, note takers who are meticulous, precise, and complete will do such things as the following:

- Explain how materials or animals were treated or handled—for example, heated, stirred, mixed, housed, fed, or tested.
- Note calibrations and calibration history of instruments to affirm that they are functioning properly.
- Record relevant time frames in procedures or in observations, such as reaction rates, changes in color or behavior, or duration of observation.
- Describe techniques used to purify or test any starting materials or reagents, and show the results.
- Show all details of mathematical calculations or statistical applications.
- Use correct names for laboratory glassware and vessels; for instance, was the material placed in a cylinder, flask, pipette, crucible, dish, or beaker?

Any number of situation-specific items could be added to this list, such as those associated with particular instruments, techniques, experimental conditions, mathematical expressions, statistical applications, or the definition of unique terms and units. Each note taker must assess the recording needs for both the given experiment and the notebook's readers.

Experimental notes also may incorporate information in visual forms—such as tables, graphs, flow charts, diagrams, photos, or instrumental outputs—that are either sketched, attached, or (least preferably) kept in a separate but carefully cross-referenced location such as a labeled folder or another notebook used just for such purposes. In the alcohol project example, the observed differences in average drinking rates of butanediols for high- and low-selecting mouse strains, such as the "drinking index" mentioned in Ex. 2.6, could be drawn as a bar graph (Figure 2.1).

When using a visual representation of collected data, whether hand-drawn or electronically generated, it is helpful to label it with a number and experimental title (for cross-referencing) and to include an explanatory caption with the date(s) the data were collected. If not already indicated, the caption may also provide a key for reading the visual correctly (e.g., axes, columns and rows, symbols, or colors) or giving the data context (e.g., standard error or deviation ranges). As shown in Figure 2.1, keep visuals simple and readable: a series of simpler drawings or graphs is easier to follow than a single figure that is comprehensive and overloaded with information. A great benefit of visual representations is that they can *show* much information in a more condensed way than writing out the same details. A fully useful visual, especially when looking back later, also is accompanied by the appropriate notations for seeing its significance correctly and unambiguously. For instance, the usefulness of drawings, whether of objects, phenomena, or behaviors, is enhanced when the notes provide such associated information as scale, dimension, intensity, motion, material composition, physical qualities, and details known through other senses, such as tactile or olfactory. Near the visual, also note the location of original data on which it is based.

Broadly speaking, avoid any infusion of ambiguity into the notes. All recorded observations, whether next to visuals, on special data forms, or directly in your narrative, should be highly legible. Letters and numbers should not be subject to misreading. Moreover, an illegible record might result in wastefully having to repeat experimental work. Another potential source of ambiguity in an experimental narrative is a grammatical one: a misuse of tense

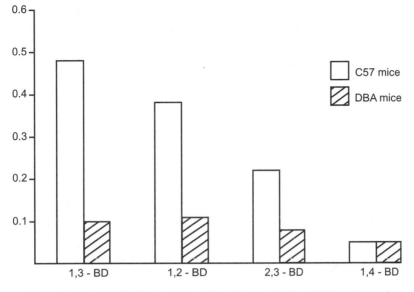

Figure 2.1 Alcohol drinking index for 15 high-drinking (C57) mice and 15 low-drinking (DBA) mice, showing four types of butanediols (along the x-axis) and consumption rates (y-axis) over a 10-day period (September 15–24, 1974)

that blurs the distinction between fact and expectation, such as actual actions and observations versus intentions and speculations, which belong in the introductory notes. Thus, while introductory notes may contain anticipatory phrasing—"the intent is" or "it is expected that" or "this should then," for example—the narrative is written in the past tense: "butanediol *was* consumed," "the animals *did* not," or "we *injected*." An otherwise sound record should not be rendered unreliable by readily controllable factors like penmanship, grammar, or precise usage of words.

How well and how thoroughly your notes tell the story of your experimental work will become evident whenever they are consulted later, or used for laboratory reports, oral presentations, or professional papers. The fullness of a notebook's record is what makes it dependable for subsequent and even unanticipated uses. That is why notes should contain every measurement taken (e.g., weighings, sensory readings, absorption intensities) and all calculations used to convert *raw data* (such as how much fluid an animal drinks) into *derived data* (a drinking index). The fact that only derived data typically appear in a report makes preserving the raw data even more essential. Should logical fallacies become apparent later, for instance, recalculations will require the original measurements. Or, if an instrument is discovered to have given false readings, access to the original data will allow the application of correction factors and reinterpretation of what actually happened. Simply put, there is no substitute for a clear and scrupulously thorough record. The raw or primary data collected—from actual measurements of such variables as volume, mass, time, and intensity—constitute a precious resource that underlies any subsequent thinking and writing associated with the research.

Discussion and Conclusions

Once recording what was done and what occurred is completed, a clear transition or a heading demarcation should be used to indicate that the remainder is *reflection* on the results. This final component of thorough notes looks back at what happened and offers interpretations, suggests practical or theoretical implications, points out experimental limitations, makes conclusions, and even looks forward by suggesting further experiments. Of primary concern initially is to return to the hypothesis. In our alcohol project example, the notes must respond to the key question: Do the results support a neural sensitivity theory of drinking differences in laboratory mice? How strongly? Are there uncertainties, reservations, or qualifications? Here is how part of that discussion might proceed:

Ex. 2.8

Since three of the four butanediols tested (1,3-, 1,2-, and 2,3-BD) show differences in drinking and neural sensitivity parallel to ethanol, the results lend support to the role of inherited brain differences. As to 1,4-BD, which is almost totally avoided by both high and low drinkers, other than possible toxicity I can't explain why they won't drink it, even though it has similar biphasic effects to the other alcohols. Interesting anomaly. Are smell or taste factors? (unlikely, but recheck Rodgers 1972). Maybe the best route is to compare effects on neuronal chemistry? Follow this up.

This section of the notes should be a freewheeling and unfettered consideration of the findings that may include calculations, drawings, and scattered musings that speculate, synthesize, make connections (including bibliographic), and in the luckiest of circumstances capture a surprising insight that might otherwise have gotten away. Beyond the conclusions associated with your hypothesis, a forward-looking discussion also addresses how the insights from these results may be of practical utility, either more narrowly within the field or perhaps commercially.

Well-organized notes that contain painstaking procedural and observational details, comprehensive discussion, and carefully derived conclusions are invaluable. For one, solid notes are much easier to use for writing laboratory reports or articles. While theoretical arguments or speculative conclusions may be the more memorable aspects of papers, it is the carefully recorded experimental evidence that gives them scientific value. A meaningful and reliable account of experimental work can come only from a good set of records—on-the-spot, objective, clear, detailed, comprehensive, accurate, and thoughtful descriptions—together with instrumental outputs. As evident as it may seem, notebooks should not be discarded. Scientists departing from research groups typically leave their notebooks behind for access by co-workers to experimental details available nowhere else. It is wise to keep copies for personal reference, unless prohibited by proprietary rights like those in industrial or government research settings. In sum, the note-taking habits that one develops and maintains will determine the ultimate usefulness and accessibility of the record, and reflect the value placed on record-keeping basics.

ELECTRONIC NOTE TAKING

Although electronic notebooks are an option available commercially, for the most part laboratory notes continue to be taken in the time-tested way, by hand. Computerized notes can pose such problems as keeping them secure, unalterable, reliably dated, and authentically witnessed, all of which have legal implications. Printing out electronic records, along with any subsequent additions or revisions, is a way to enhance their authenticity and allow for hand-dating and witness signatures. Such hard copies then may be attached permanently to the pages of a bound notebook, as well as stored by an institutional archivist who can attest to their integrity. In addition, electronic records must employ a recording and storage system that renders notes unalterable. For this purpose, one recommended software is Write Once, Read Many (or WORM).[9] An example of currently available software for maintaining an electronic notebook is LabTrack, advertised by Avatar Consulting in Laguna Hills, California, as a "Legal Electronic Lab Notebook." Electronic notes may also be taken using speech recognition systems that will convert your dictation into text. One such system is Byblos, marketed by BBN Technologies of Cambridge, Massachusetts, as the "BBN Hark Recognizer."

Legal considerations aside, electronic notes do have some advantages over handwritten notes in that they:

- are more legible;
- facilitate the conversion of data into visual forms;
- provide quick and easy access at all times;
- can be readily shared through e-mailing;
- are protected from chemical (or coffee) spills.

The process of writing on a computer, however, is less spontaneous and slower than handwriting, so in effect it is less natural for the immediacy of laboratory work. Although electronic speech recognition technologies are advancing rapidly, for the purpose of taking laboratory notes the risk of errors in voice-to-print conversion must be perceived by the note taker as negligible or inconsequential—a high bar—before these systems can be accepted as completely trustworthy.[10] Time, practical experience, and legal precedents in the making will test the usefulness of such technological developments. Even assuming all the legal issues can be fully resolved, it still remains to be seen whether electronic note taking will grow in favor.

LABORATORY REPORTS

Notebooks that adhere rigorously to the professional expectations described here will facilitate the writing of subsequent documents based on them, such as laboratory reports, graduate theses, journal articles, grant reports, and oral presentations. Of interest here is the write-up of the notebook entries into a laboratory report, whether for a periodic progress report at a biotechnology company or an assigned report in a college laboratory course. Aside from workplace variations in such aspects as format or witnessing, two basic differences between notebook entries and laboratory reports are their level of formality and extent of detail. First, the informal and often clipped or abbreviated phrasing in notes must be converted grammatically and stylistically into full scientific statements, explicitly interconnected, that officially communicate the work to administrators or instructors. Here, the shared conventions that ensure clarity and precision in scientific English become all-important. Second, the comprehensively detailed notes must be reduced to the key information that will suffice for understanding, repeating, and validating the work. There are various kinds of notations made during an experiment— trial runs, personal reminders, names of standard materials or instruments (as seen in Ex. 2.5)—that would clutter up a report with unnecessary details. Working procedures and informal notations must be distinguished from the necessary reportorial elements. This example illustrates original lab notes, followed by a second version where they are streamlined for a report:

Ex. 2.9
1. Centrifuged liver homogenate for 20 min at 270g (1500 rpm, SS-34 rotor, 4.25″ radius, using Sorvall Superspeed RC2-B Automatic Refrigerated Centrifuge, 6°C. Kept supernatant (brownish) and discarded pellet (yellowish, reddish). Homogenize a little longer next time, see what happens. Store homogenate supernatant in refrig until Biuret reagent reaction. Complete standard curve.
2. The homogenate of each liver was centrifuged at 6°C for 20 min at 270 × g and the supernatant immediately assayed for NAD reduction.

Selectivity in giving information does not of course mean misrepresenting in any way what actually occurred experimentally. Maintaining professional trust among researchers does mean that under no circumstances should data be reported misleadingly or laboratory notes altered to accord with an outcome presented in a report. Whatever the differences among forms of scientific writing, a common denominator must be complete accuracy and truthfulness.

Beyond situation-specific practices regarding form or content, a laboratory report contains, like notes, the standard components of the IMRAD model (introduction, methods, results, and discussion), which is widely followed in the experimental sciences, especially in chemical and biological research. The difference between lab notes and a lab report is one of selectivity, formality, and critical thought. For instance, a laboratory report's conclusion discusses more comprehensively how the experiment went: what procedural details need refinement, how well the hypothesis was supported, methodological limitations, and new hypotheses suggesting further experiments. The IMRAD model is adapted for application in the technologies, physics, and engineering, in accord with the kinds of purposes, methods, outcomes, and stylistic conventions that meet the communication needs in these fields. The following numerically sectioned outline, for the alcohol project in Exs. 2.1–2.8, illustrates the conventional structure and content of a laboratory report.

Ex. 2.10

1 INTRODUCTION

  1.1 Historical background and purpose of current project

    1.1.1 Identification of drinker and non-drinker laboratory mice

    1.1.2 Biochemical mechanisms underlying alcohol drinking in mice

    1.1.3 Measurement of drinking, sensitivity, and metabolism in mice

  1.2 Rationale and sources for methods used to measure consumption, sensitivity, and metabolism using butanediols

2 EXPERIMENTAL PROCEDURE

  2.1 Materials

    2.1.1 Animals (mice, housing, food, suppliers)

    2.1.2 Chemicals (butanediols, reagents, sources)

    2.1.3 Equipment (activity box, centrifuge, spectrophotometers)

  2.2 Description of the work

    2.2.1 Measurement of consumption of 1,2-, 1,3-, 2,3-, and 1,4-butanediol

    2.2.2 Tests for neural sensitivity (open-field activity, jaw-jerk response)

    2.2.3 Assays for butanediol metabolism (NAD reduction)

3 RESULTS

  3.1 Consumption indices for butanediols

  3.2 Neural effects of butanediols on neuromuscular activity

  3.3 Rates of butanediol metabolism in liver homogenates

4 CONCLUSIONS AND DISCUSSION

  4.1 Similarity in outcomes between 1,3-butanediol and ethanol

  4.2 Rejection of 1,4-butanediol by all mice

  4.3 Support for neural and hepatic determinants of drinking behavior

5 REFERENCES

At this more formal level of writing, typically meant for supervisory readers (removed from the project to various degrees), it is important to be meticulously explicit both in describing the work and in articulating connections to its broader scientific context. Of particular use to readers who are not close to the work, beyond the methodological details and results, is the discussion and conclusions section. To what extent were the hypotheses supported or rejected? Were there any unexpected or surprising results? Were there anomalies in the results that suggest further experiments or new approaches? Do such unexpected results suggest important implications or applications regarding the phenomena studied? Did the equipment or procedures present measurement limitations? Could any such limitations be addressed with access to or acquisition of other equipment, or by certain procedural changes? As a first and immediate approximation of what occurred experimentally, a laboratory report is still another intermediate document that may be used to write the most formal or official public documents, such as articles and grant proposals. A laboratory report exhibits senses in which the work is still in progress, with its formality tailored for internal purposes such as workplace updates or instructional assignments. Some of the language or local references may reflect those internal purposes and some visuals may be hand-drawn rather than computer-generated. Therefore, a laboratory report may be viewed as a full but initial verbal crystallization, a first-order on-the-scene narrative limited to local purposes and readers.

FROM LABORATORY RECORDS TO OTHER COMMUNICATIONS

When the information extracted from laboratory notes and reports is used for more sophisticated documents, the expectations of the intended readers together with the required artifices of formality will determine how they are written. For instance, the features of language, form, and content that are required in writing a federal grant application are readily distinguishable from those expected in a journal article. As the most developed statements to peers in the profession, publications like journal articles and book-length monographs represent the highest-order synthesis of experimental work and thought in a field. Scholarly writing, whether in an article or in a college report based solely on bibliographic research, goes beyond the scope of a laboratory report in significant ways. For one, scholarship typically encompasses and connects with the larger body of work and theory in that experimental niche (e.g., alcohol studies with mice, gene mapping in tomatoes). Second, it is a more thorough, thoughtful, and persuasive presentation of the theoretical and practical aspects of the work being shared. Contrary to a tempting but misguided assertion, the data do not speak for themselves. The observed outcomes must be subjected to careful interpretation and the conclusions should spark attention to the broader scientific implications: Has the work discovered a geophysical phenomenon, revealed a biochemical effect, refined a treatment modality, or extended a theoretical perspective? There are evident and necessary differences in formality, sophistication, and comprehensiveness across the various forms of scientific writing, from the immediacy of laboratory notes to highly developed scholarly writing. The one consistent thread that runs through the scientific record is the IMRAD model of writing that suits the peculiar needs of experimental inquiry. Although the IMRAD model may not fit the requirements of note taking in every scientific discipline, in biology and chemistry particularly it is standard practice. Those who submit experimental papers for publication are expected to adhere to the IMRAD method of organization in reporting their work.

The process of taking laboratory notes underscores the primary importance of the notes as the basis for the authority, credibility, and usefulness of subsequent and more highly formalized research communications. In the middle ground between laboratory writing and scholarly publication, there is a range of functional, routine, and relatively short workplace communications of varying formality and purpose. These include abstracts for conference papers, letters to peer researchers, and internal memoranda for conducting daily business in scientific occupations.

WORKPLACE SCIENTIFIC WRITING:

LETTERS, MEMORANDA, AND ABSTRACTS

THE ROLES OF WORKPLACE SCIENTIFIC WRITING

Between the immediacy of experimental writing (notes, lab reports) and the more formal communication of research (student reports, articles), there are numerous routine forms of writing used in scientific job settings. The importance of these everyday communications is easily overlooked or taken for granted, but they constitute the administrative glue of workplaces. The daily professional responsibilities in scientific organizations require the writing of documents that officiate, organize, and conduct scientific business. Although recording and publishing research are key forms of scientific writing, such workplace documents as memoranda and letters are also significant. They allow a system of close and documented communication, as well as an official day-to-day paper trail, without which organized scientific endeavor would rapidly falter. There are also short scholarly forms, such as abstracts, article reprint requests, and notes or letters in periodicals. Less frequently, scientists write for public media (as in press releases, opinion letters, news articles). These basic forms of written communication in the scientific work world—memoranda, letters, abstracts, and public exchanges—are adapted to suit a wide variety of recipients and purposes; a number of these various uses are listed in Table 3.1. When employers hire scientists—whether in corporate,

Table 3.1 Types and purposes of routine workplace communications

Document Type	Sample Purposes
Memoranda from administrators	Explain a new workplace policy; announce arrival of a new employee; evaluate employee achievement periodically; summarize annual institutional activity
Memoranda from employees	Request supplies; report on business trips; summarize professional activities periodically
Letters from job applicants	Highlight career qualifications, with résumé and other enclosures, such as publication copies
Letters to job applicants	Inform about hiring decisions (with either good news or bad news approaches)
Letters to clients	Offer scientific services (e.g., soil testing; genetic analyses); advertise new products; report annual earnings
Letters to editors	Comment on scientific technicalities or issues to peers in journals or to the general public in newspapers
Letters from scientist-editors	Convey comments or decisions on submitted manuscripts; invite manuscripts for special journal issues or book series
Letters among colleagues	Request article reprints; inform on political issues; inquire about experimental techniques; discuss scientific ideas
Letters of support	Recommend candidates for jobs, awards, and other duties or honors
Press releases	Announce discoveries; comment officially on issues; inform citizens of public hazards (e.g., environmental, biochemical, dietary, pharmaceutical)
Letters to public and legal officials	Support scientific initiatives (e.g., funding, legislation, policy); provide testimonials (hearings, court cases)
Abstracts	Summarize articles, oral presentations, formal institutional reports

academic, or government settings—they trust that practical, on-the-job documents will be written effectively.

Job application letters, résumés, inquiry letters, reprint requests, progress memoranda, and research abstracts are universal forms having wide application at all levels of the scientific community. These forms can also be incorporated in practical ways into undergraduate scientific coursework. Just as a corporate scientist may periodically submit a memorandum report that summarizes progress on a particular project, for instance, students similarly can write progress memoranda on an experiment or a research paper to an instructor as well as to fellow students with whom they may be working collaboratively.

JOB APPLICATION LETTERS AND RÉSUMÉS

Job applicants typically submit two basic items: (1) a cover letter that formally states their intent to apply for the opening and that highlights their qualifications and career objectives, and (2) an attached listing of biographical information, especially education and relevant experience, commonly called a résumé or a curriculum vitae (CV). The latter is sometimes more narrowly defined as containing only one's academic and professional achievements, without an employment history. Applying for advanced positions may require additional items such as publication samples or (given the expense of doing science) evidence of successful grant-supported work. A complete application may ensure full consideration, but candidates who make the short list for serious consideration will exhibit both their scientific qualifications and their technical writing competence. Given that the data provided on résumés (no less than in lab notes) do not speak for themselves, a job letter is the place to speak up on one's own behalf. The levels of training and experience that applicants use to build their appeal for an interview will vary, but the letters of all applicants reveal their writing ability—how well they use language, organize information, and convey facts with technical rigor. Scientific knowledge and experimental outcomes that cannot be conveyed effectively in standard professional ways are of limited use. The cover letter itself, therefore, becomes all-important in its own right as a sample of an applicant's writing. Although the résumé naturally must be prepared first, technically it is an attachment to the applicant's personal appeal in the letter's introductory narrative.

JOB LETTERS: HIGHLIGHTING PROFESSIONAL KNOWLEDGE
AND WRITING ABILITY

What are the basic features of a well-prepared job application letter? As with other kinds of writing, there are certain expectations the writer must meet regarding details of form, content, and readability. Job applicants must observe these standards in two regards: First, and more broadly, they must show competence in following common work-world practices—such as in giving dates, addresses, salutations, and signatures, and in using a business diction. Second, they must meet field-specific norms that include competent use of technical language and a coherent recitation of professional qualifications. There is no single correct way for a job letter to sound, and in any case every letter carries a writer's individual voice, which should at the very least be confident, positive, unaffected, and appropriately deferential without either understating or over-pitching. The following hypothetical letter is appropriate for a fresh graduate with a Master of Science in biology.

Ex. 3.1
937 Orchard Lane
Indiana, PA 45701
April 25, 1975

W. S. Carlton, PhD, Director
Behavioral Genetics Program
University of La Jolla
La Jolla, CA 10791

Dear Dr. Carlton:

    I am writing to apply for the Research Associate opening in your Behavioral Genetics Laboratory, posted in the April 18 issue of Alcohol Studies Quarterly (Ref #507). My graduate research is on inherited differences in alcohol drinking behavior in mice as a model for understanding the metabolic and neural factors involved. These interests fit well with your program's aims.

    On April 8, I defended my thesis on neural sensitivity of lab mice to butanediols for my MS in biology at IUP. On April 14, I delivered a paper on my findings at a meeting of the Federation of American Societies for Experimental Biology (FASEB), in Atlantic City, NJ. Two articles on this work will

---

appear in the May and September 1976 issues of *Biochemistry, Pharmacology, and Behavior* (preprints enclosed). Though I plan to pursue a PhD in biology, currently I seek research experience in a team-oriented academic setting and hope later to integrate work with school.

Animal behavior has been a passion of mine since an unusual summer volunteer experience in the immunology lab of Dr. Edward Boyce at the Sloan Kettering Cancer Center in NYC, just before starting college in 1967. In Stony Brook's biology program (BS, 1971), this interest was solidified with such courses as Animal Behavior, Animal Learning, Neurophysiology, Field and Theoretical Ecology, and Non-Human Primate Ethology. I have taught general and cell biology labs as a graduate Teaching Assistant and would enjoy teaching introductory biology and animal behavior courses at ULJ.

Thank you for reviewing my attached CV for a research position in your program. If my background meets your needs, I would welcome an interview at your convenience.

Sincerely,

Robert C. Goldbort

Enclosures: CV; FASEB abstract; article preprints

This letter contains the various elements of style, content, and organization that a job applicant is expected to include, from the necessary addresses down to the list of attachments. The letter's four-paragraph narrative progresses as follows:

- *Paragraph 1:* States directly and concisely the intent to apply for a specific opening, underscoring key qualifications (education, research) as a transition to the details.
- *Paragraph 2:* Highlights relevant educational and professional details (thesis defense, publications, career goals).
- *Paragraph 3:* Demonstrates a longstanding personal interest in the field (pivotal volunteer experience, relevant undergraduate courses).
- *Paragraph 4:* Closes by expressing appreciation for being considered and interest in an interview.

With both letters and résumés, there is always the question of how much detail to include and whether to heed the commonly dispensed advice—really a myth—to keep each to one page. One applicant may barely eke out a full page while another may struggle with the opposite problem of restraint. Each applicant must strike a personal balance between demonstrating the advertised requirements and including less consequential details that may best be summarized or left out. Even when a résumé is inclusive and lists items not directly career-related, the writer may focus on selected items in the letter.

Decisions about what to mention and what wording to use are a common concern for undergraduates, whose work experience in particular tends to be very limited or not directly relevant to their career goals. When work experiences are not directly career-related, they can be mentioned in the letter if the duties involved have transferable value, such as multitasking, handling financial transactions dependably, supervising or training employees, managing time efficiently, adapting quickly to new situations, working independently, and being innovative. Students who have no employment history can focus more extensively on such areas as their course experiences, independent research with faculty, the types of experimental equipment they have learned to use, internships, specific career goals, and any plans for further education. Students applying for an internship as their senior year approaches can include similar kinds of content in the cover letter. A common pitfall in student letters is writing about qualifications in generalities, with vague or abstract sentences that lack supporting examples. When computer literacy or independent experimental work is mentioned, for instance, such items become clear and real when accompanied by the names of specific software or details of experimental methods and goals.

Finally, one must remember that the language in a letter does not just provide information but also exhibits the writer's personality and attitude. Do the word choices and phrasing appear to inflate the applicant's qualifications (e.g., "unmatchable" experience, "tremendous" drive, "vast" knowledge)? Does the language show more concern with personal gain from the position rather than with offering specific assets to the organization? A "you" viewpoint (as it is sometimes called), versus a "me" orientation, emphasizes the interests of the reader—in this case, the person who may be deciding whether to grant an interview. One way to be you-oriented is to use the words "you" and "your" more frequently than "I," "my," "mine," and "me," especially at the beginnings of paragraphs and sentences, places of natural emphasis. Another way is to show specific knowledge about the organization that makes the position personally appealing. In any case, an effective letter appeals coherently, concretely, and convincingly for full and serious consideration. It highlights the applicant's key qualifications and sets up anticipation for reviewing the attached résumé.

RÉSUMÉS: LAYING OUT THE FACTS

Unlike a letter's narrative form, with full sentences and a personal voice, the résumé is a matter-of-fact, abbreviated listing of biographical information. Our era of word processing facilitates an array of design and layout choices that can be as bold and innovative as an applicant may wish to risk. At the same time, however, the visual aspects must be selected wisely to enhance rather than to impede readability and appeal. Constructing a résumé calls for various kinds of decisions, including options regarding the following:

- Personal information
- Type and order of categories
- Selection of content details
- Layout and design
- Typographical features
- Paper size, texture, and color
- Language and phrasing
- Length

Due to features of design, layout, typography, and sometimes color, résumés are visually more dynamic than letters. Their typical organization is chronological, with inverted date order, as in this one (accompanying the letter in Ex. 3.1).

Ex. 3.2

Robert C. Goldbort

937 Orchard Lane, Indiana, PA 45701

(412) 314-1953

EDUCATION

1975   MS, Biology, Indiana University of Pennsylvania, Indiana, PA

1971   BS, Biology, State University of New York, Stony Brook, NY

TEACHING EXPERIENCE

1973–1975   <u>Graduate Teaching Assistant,</u> Biology Department, Indiana University of Pennsylvania, Indiana, PA. Courses taught:

	• General Biology Lab
	• Cell Biology Lab
1971–1972	Assistant to the Director, Composition Program, Queens-
	borough Community College, Bayside, Queens, NY:
	• Scheduled and supervised peer composition tutors
	• Tutored writing individually and in small groups

PUBLICATIONS

Strange, A., Schneider, C. W., & Goldbort, R. (1976). Selection of C_3 alcohols by high- and low-ethanol-selecting mouse strains and the effects on open-field activity. *Pharmacology, Biochemistry, and Behavior,* 4 (5), 527–530.

Goldbort, R., Schneider, C. W., & Hartline, R. (1976). Butanediols: Selection, open-field activity, and NAD reduction by liver extracts in inbred mouse strains. *Pharmacology, Biochemistry, and Behavior,* 5 (3), 263–268.

Goldbort, R. C. (1975). *A study of the butanediols as an approach to understanding the relationship of alcohol tolerance to alcohol preference in inbred strains of mice.* MS thesis, Indiana University of Pennsylvania, Indiana, PA.

Goldbort, R. (1975). Selection of butanediols by inbred mouse strains: Differences in specific activity and central nervous system sensitivity. *Federation Proceedings,* 34 (3), 720. Fifty-Ninth Annual Meeting of Federation of American Societies for Experimental Biology, Atlantic City, NJ.

REFERENCES

Provided upon request

Besides the three major categories in this example (education, teaching, publications), the individual experiences and qualifications of applicants may call for other categories, such as grants awarded, academic and professional honors, service and volunteer activities, military duty and special training, and special skills like computer programming or foreign language proficiencies. Alternatively, some may prefer a *functional* organization, categorized by marketable skills and experiences that the applicant may wish to highlight, such as supervising, training, managing, grant writing, or consulting. Whether chronological or functional, or some combination, a résumé commonly is headed by the applicant's name, addresses and phone numbers (work and home), a fax number, and an e-mail address. As to other personal information, with few and justifiable exceptions one is not legally required to provide certain details, such as age, sexual orientation, religion, country of origin, race, or marital status, although including them is the individual applicant's choice. Immediately following the standard heading and preceding the major categories, applicants may opt to include a line or two stating their job objective (e.g., "Seek a research associate position in a team-oriented corporate setting with opportunity for advancement to management"). Once the major categories are determined, one must make decisions regarding the specific entries under each of those categories, the degree of detail provided, and the style and layout of the information. Content considerations may prompt such questions as: Should experience, skills, or interests be listed that are unrelated to the desired work? Which duties or achievements should be included under each entry? Should a particular work experience be omitted if it leaves a noticeable gap in the chronology? Whatever information one does decide to include must be presented accurately and ethically, without any intent to mislead or misrepresent.

Choices regarding the design, layout, and typography must work in favor of, rather than against, the flow of the information or the perception of the applicant's personal qualities. To begin with, opting for the convenience of a software template for a résumé may risk a perception of unoriginality or laziness, besides the possibility that the templated design itself may not be suitable or appealing. Such templates usually are available with word-processing software, such as Microsoft Word, or can be purchased as a separate package. Applicants should make their own stylistic choices using the many features available in conventional word processors, and without drawing attention to them for their own sake. For instance, avoid using too many different fonts, letter sizes, or bullet styles; distracting boxes, internal and border lines, or color-coding of headings; excessive bolding and italics; or an unnecessary series of indentations (versus block style). In the résumé shown above, a few simple design and stylistic features are used to enhance readability: double-spacing between entries, all capital letters for headings, italicized titles, simple bullets, and consistent left-hand positioning of dates. Prospective employers will be grateful for the efficiency of short phrases in bulleted lists over having to wade through a prose style that belongs in the letter. For those who may wish to experiment with the extreme end of design options, specialized software can be used to create all-graphical résumés or to turn a list of qualifications into a "billboard of achievement" or, more radically, a "baseball card résumé" with the applicant's photo accompanied by his qualifications listed like a player's statistics.[1] However, even the conservative features of layout and typography in Ex. 3.2 will allow plenty of stylistic latitude without necessitating visual drama or a gimmicky ad-campaign approach that will likely distract readers from the substance of the résumé. Features of layout and design may be already templated for job seekers posting their résumé with online search services (such as monster.com) to which companies may subscribe.

The choices in application materials, involving everything from content and language to format and length, are the applicant's alone to make. Among the less weighty aspects to consider are the length of the materials and the color of the paper. Some may agonize over paper color, but a conservative white or cream is the least risky and most common. There are no firm rules regarding how long either the cover letter or the résumé should be. Letters typically are one to two pages long, but the extent of an applicant's background or the requirements of a particular position may call for a longer statement. Similarly, a one-page résumé may suffice for an applicant who is at the start of a career, but an experienced scientist likely will need several pages to provide an employment history, research accomplishments (including grants awarded), and a list of publications. Both the degree of detail and the résumé's style (spacing, layout) will affect its length. Considerations of length, paper quality, or color should not obscure the central concerns, namely, the integrity of the application process and the writing of materials that will persuade a prospective employer to grant an interview and ultimately offer the job in the competition with an unknown pool of applicants.

INQUIRY LETTERS

Besides job application letters and résumés, another common type of short communication within the research community is the inquiry letter. One simple variety of this type is the reprint request. Researchers use such requests to inquire about one another's work by asking whether an article reprint (a separated offprint from the original journal issue) is available. The form of the request may be a traditional letter, a post card, or electronic mail. Reprint requests are brief and highly formalized (or templated), like this post card in English, French, and Spanish used by a researcher at a Venezuelan university who just types or writes in the citation and signs the card.

Ex. 3.3
Dear Dr.
Monsieur Le Professeur Dr.
Estimado Doctor

I would greatly appreciate a copy of your paper.
Voudriez vous avoir l'obligeance de m'envoyer un exemplaire de votre article.
Agradeceriamos a Ud. Una separata de su trabajo.

```
R. Goldbort
Ethics in Scientific Writing, Journal of Environmental
Health, 55/2, Sep-Oct 1992, 52-53.
```

Sincerely yours,
Remerciements anticipés,
De Ud. Muy atentamente,

Dr. C. Cressa
Universidad Central de Venezuela
Institutio De Zoologia Tropical
Caracas, Venezuela[2]

For the recipient's convenience in responding, such post cards often carry a peel-off return address label. While post-card inquiries are a simple and convenient way for scientists to exchange publications, more involved inquiries necessitate a conventional letter format or, if signature is not an issue, electronic mail. Using e-mail permits rapid exchanges of information that can include file attachments or relevant online links. In the United States, the advent of online databases has provided access to articles in PDF form and minimized the practice of making reprint requests.

The following electronically sent letter, in this case rather formal and highly formatted for readability, inquires about the effectiveness of software used for teaching.

Ex. 3.4

From: "Robert Goldbort" (ejgold@root.indstate.edu)
To: bryanjg@ucenglish.mcm.uc.edu
Subject: Daedalus for tech writing
Date: Thu, 9 Feb 1995 17:28:24

Dear Professor Bryan:

I've read the article in the *Chronicle of Higher Education* about your use of Daedalus with technical writing students. I teach tech writing in the English Department at Indiana SU and wish to experiment with conferencing software to teach at a distance. I'm curious about the following:

1. **How much time, trouble, money, and other resources does it take to set up a technical writing class using Daedalus?** How much equipment and space? Do you create special handouts or instructions to distribute to the enrollees?

2. **Are administrators supportive?** Do they see distance courses as important? Do they act eagerly and promptly to assist such teaching efforts? (e.g., any grants?)

3. **Is Daedalus among the best conferencing programs for distance teaching?** How does it compare with other available teaching software? For instance, are you familiar with CoSy 5.0 Groupware Environment?

4. **Does Daedalus permit your students to send you technical reports with graphics?** Pegasus Mail has difficulty sending graphics, so I wondered whether Daedalus permits users to attach files with charts or scanned images?

5. **Do you get comments from students as to how they like the process?** What they like most? Least? Do they groan or drop the course due to the challenges of electronic learning, such as using the software? Are your office hours online?

6. **How far away from campus are your students located?** Are they mostly working students taking a few credits? Do they use computers at home or at work to participate in discussion? Do they fax assignments or just e-mail them?

**7. Is your distance technical writing course listed in your college cata-
logue or registration materials and open for anyone to take?** Do you specially recruit students for the class? Is there a special Web site listing your
class?

Thanks for any information/opinions/advice you could offer to a distance
teaching neophyte hoping for a successful first try. I plan to order group software by mid-April.

Sincerely,
Rob Goldbort, PhD
Associate Professor of English
Indiana State University, Terre Haute, IN 47809[3]

Electronic correspondence can be formal or informal and is easy to manage,
especially for prioritizing and archiving or for multiple mailings using distribution lists. Since your inquiry's recipient may not be obligated to respond, it
should be written with that individual's convenience in mind. To that end, the
inquiry example above enumerates and spaces the questions so they are readily distinguished, and emphasizes each main question preceding the more detailed follow-ups. Naturally, any response will be facilitated to the extent that
the writer provides a clear sense of the inquiry's purpose together with specificity and clarity of the questions themselves.

Scientists also may make inquiries in more public roles that include policy
advocacy. As members of private special-interest organizations that may ad-
vocate and support particular scientific goals or policies, scientists may write
letters to colleagues, government officials, or citizens to request their support
and participation. Such letters may include research inquiries in the form of
opinion surveys or questionnaires. The astrophysicist Carl Sagan, for example, as president of the Planetary Society sent a two-page letter in the early
1990s—addressed to "Dear Fellow Citizen of Planet Earth"—outlining the
society's aims and asking readers to fill out an enclosed Space Policy Questionnaire regarding specific US government space initiatives. The questionnaire's nine items ranged in subject from Mars missions and NASA's budget
to the space station and SETI (search for extraterrestrial intelligence). "Your
replies," Sagan closed, "will help us influence government leaders as they
consider the national and international space agenda and expeditions to other

worlds." Or, consider an October 1991 public letter to colleagues at large from the Nobel-laureate physicist Henry W. Kendall as chairman of the board of directors of the Union of Concerned Scientists. Like Sagan, Kendall explained his organization's aims (to deemphasize military research in favor of solving "pressing environmental and social problems") and asked readers to become sponsors and to fill out an enclosed Survey of American Scientists. The survey's 14 questions covered the environment (greenhouse warming, for example), arms control (the nuclear threat), and professional issues (scientific education). Kendall's closing called on colleagues to exercise social responsibility by joining the group's Scientists Action Network "to become actively involved in our efforts to create a better world, both with regard to arms control and the environment." Letters like those of Sagan and Kendall, as well as letters to editors or to elected officials, allow students and working researchers alike to engage their broader and fuller professional responsibility as members of the scientific community. Given the considerable public funding of scientific research (in the billions annually), it behooves all researchers to understand, be able to work within, and apply critical and anticipatory thinking to the national debates over science policy and its long-term planning. As Kendall notes in his letter to colleagues: "What we do affects the lives of billions of people outside our laboratories, both in our own time and for generations to come."[4]

Besides inquiries and job letters, scientists may write all sorts of other letters—traditional or electronic, formal or informal, public or private—ranging from personal communications with colleagues to letters associated with employees or students and published letters in research periodicals. Workplace communications like letters and memos may seem extraneous to research, but they are nonetheless an important part of the organizational and social fabric of the scientific professions.

TECHNICAL MEMORANDA

Unlike letters, memos are written for internal readers. However, just as with letters, scientists may write memos for various on-the-job purposes. In administrative capacities, scientists may need to announce and explain to employees (technicians, researchers) new workplace practices or policies, including those relating to experimental protocol. Administrative memos also may share information regarding budgets, grant activities, or corporate profits. The official paper trail moves in the opposite direction as well, with employees sometimes using memos to provide information to supervisors. A memo may be as simple as a transmittal note that explains an attached document or a request for lab materials or funding for a trip, or it may be a more involved statement regarding professional activities as part of annual employee reviews. Here is a hypothetical example that illustrates how letters and memos differ as business forms.

Ex. 3.5

GENETIC APPLICATION TECHNOLOGIES

Interoffice Communication

To: All Research Associates
From: Dr. Karl Robertson, VP for Research and Development
Date: September 3, 2003
Subject: Company Policy Updates

Please be aware of the following revisions to company policy in the research division, approved by GAT's Board of Directors and effective immediately:

1. *Travel approvals:* Use the new travel forms, and provide specific details for the added questions on foreign travel (priority, length of stay, anticipated benefits).

2. *Materials requisition:* The revised form for ordering lab supplies requires Director and VP approvals, so please anticipate an extra day for full processing.

3. *Archiving lab notebooks:* Note that archiving completed lab notebooks now requires the signatures of two associates and the research VP besides the Director's.

Thank you for your cooperation regarding these important procedural updates.

Attachments: Revised travel, requisition, and archiving forms
Cc: Dr. Sarah Jensen, CEO; Jonathan Sanders, Company Attorney; Research Directors

---

As vehicles for conducting internal affairs from day to day, memos are direct, functional, and addressed to a limited and familiar audience. In place of addresses and a formal salutation, there are simple "to," "from," "date," and "subject" (or "regarding") fields, and contrived introductions or closings typically give way to a straightforward manner. Like letters, however, memos vary in purpose, formality, and style. A memo that proposes new initiatives, explains their potential value, and appeals to employees for their support and suggestions will address its readers differently from the example shown above. As in letters, the writer naturally will use a tone that suits the purpose for so-called good news and bad news situations.

While Ex. 3.5 is a common type of administrative memo—in this case a policy update—employees in research settings may write memos as part of their own duties, one common type being periodic summaries of professional activities. This may be an annual memo to a department chair in a university, a quarterly memo to a research administrator in a biotechnology company, or a biannual grant report to a government agency. The focus of this type of memo is on describing research accomplishments since the last update period, including experimental outcomes and publication activity (also see Chapter 2's discussion of progress reports). Given the importance of such memos in the workplace, college students may learn the form in specifically applied and adapted ways—as part of the process of writing individual or collaborative scientific reports, for example, or for reporting research progress in the course of an independent study in a faculty member's laboratory. Before submitting a full draft of a report, for instance, a memo may be used to report progress on focusing the topic and finding sources. As in the following hypothetical example, the student may include an introduction that explains the topic, a section that discusses and cites bibliographic research (here in APA style), and a closing overall assessment.

Ex. 3.6

MEMORANDUM

To: Dr. Robert Goldbort
From: Janet Smith, English 398 (Scientific Writing)
Date: October 12, 2003
Subject: Progress on research paper: A genetic basis for binge drinking?

INTRODUCTION
My research paper is on binge drinking among college students, which is alarmingly prevalent. A recent study by the Harvard School of Public Health

---

(Wechsler, 2002) found that 20% of college students binge drink frequently, with double that figure for fraternity members. My paper will describe the problem's magnitude, examine its genetic link, and suggest the use of screening tools to alleviate the problem by helping drinkers make sensible decisions.

RESEARCH COMPLETED

Using keywords like "binge drinking" and "alcohol studies," I searched the Web and some article databases (Proquest, Medline). The *Journal of Studies on Alcohol* is very helpful, and recent monographs provide solid statistical data. Here are three of my sources so far:

1. Harford, T. C., Wechsler, H., & Seibring, M. (2002). Attendance and alcohol use at parties and bars in college: A national survey of current drinkers. *Journal of Studies on Alcohol,* 63(6), 726–733.
   Statistics overall and on subgroups (Greeks, athletes, gender, race) demonstrate an upward drinking trend.

2. Murphy, B. C., Chiu, T., Harrison, M., Uddin, R. K., & Singh, S. M. (2002). Liver and brain-specific gene expression in mouse strains with variable ethanol preferences using cDNA expression arrays. *Biochemical Genetics,* 40(11–12), 395–410.
   Confirms genetic differences between drinker and non-drinker mouse strains in liver and brain activity. Explores implications for human drinking behavior.

3. Wechsler, H. (2002). *Binge drinking on America's college campuses: Findings from the Harvard School of Public Health college alcohol study.* Boston, MA: Harvard University.
   A booklet that provides detailed subgroup data. For example, it shows that among fraternity members 40% are frequent binge drinkers (FBD), 24% occasional binge drinkers (OBD), 28% non-binge-drinkers (NBD), and 8% abstainers (ABS).

CONCLUSION

Current alcohol studies point to the influence of specific gene loci (DNA sites) whose physiological regulation (liver, brain) affects drinking behavior as well. Screening tests (genetic, family) provide helpful risk assessment for individuals and for health providers.

Using a memo form to discuss progress on a writing project allows students to gain early experience with work-world writing as they begin to formulate project ideas, deal with bibliographic matters, and practice incorporating visuals into text. Even in a one-page memo like this one, the densely packed content can be partitioned for readability and much detail can be conveyed about a topic, down to the short annotations or abstracts with the preliminary sources. Abstracts themselves are among the most common and important short forms of writing by researchers.

RESEARCH ABSTRACTS

An abstract is used by researchers to summarize and sometimes (when annotated) to comment on their experimental, written, or bibliographic work. Abstracts typically range in length between 50 and 250 words. Scientists prepare abstracts of their work for various purposes, most notably to provide a nutshell rendition of a journal article, but also for conference papers, poster presentations, formal reports, proposals, graduate theses, and even lectures. A common way to classify abstracts is by whether they just *describe* the overall purpose and methods of the research presented in the document and provide a sense of its main topics, or instead *inform* readers of specific details of the research, especially the results and conclusions. Writing an abstract, whether descriptive or informative, is in principle rather straightforward but in practice something of an art that one can improve at with experience. The abstract must communicate much scientific information in a highly condensed yet specific manner.

Journal article abstracts vary in format, so it is wise to consult the style guide for a particular discipline. Style guides with specific prescriptions for article abstracts are available for such fields as astronomy, biology, chemistry, physics, geology, mathematics, medicine, microbiology, and psychology. There are also publications that contain national standards, such as the *Biosis Guide to Abstracts* and the *ANSI/NISO Guidelines for Abstracts*. Biosis is a widely used electronic indexing and searching service. The American National Standards Institute (ANSI), based in Washington, DC, provides accreditation to the National Information Standards Organization (NISO) in Bethesda, MD, for approving American standards that meet ANSI's criteria. Rather than comparing stylistic variations on the basic concept, it will be more useful to focus on the differences between descriptive and informative abstracts.

DESCRIPTIVE ABSTRACTS

A descriptive abstract, sometimes also termed a topical or indicative abstract, acts as a prose table of contents. It is written *about* the research, rather than providing the actual findings. It functions primarily to tell readers the kinds of information an article contains, focusing on the research problem and providing an abbreviated and indirect description of the methods. The sentences outline the paper's areas of information, often with verbs indicating how subjects are treated—for instance, "The various potential determinants of alcohol selection patterns are reviewed." Descriptive abstracts are common for articles that review the state of a field of research, rather than reporting an original study. Here is a descriptive abstract, less overtly outlining what the paper *does,* for an article that reviews animal models used to study the genetic basis of substance abuse.

Ex. 3.7
Behavioral and pharmacological responses of selectively bred and inbred rodent lines have been analyzed to elucidate many features of drug sensitivity and the adverse effects of drugs, the underlying mechanisms of drug tolerance and dependence, and the motivational states underlying drug reward and aversion. Genetic mapping of quantitative trait loci (QTLs) has been used to identify provisional chromosomal locations of genes influencing such pharmacological responses. Recent advances in transgenic technology, representational difference analysis, and other molecular methods now make feasible the positional cloning of QTLs that influence sensitivity to drugs of abuse. This marks a new period of synthesis in pharmacogenetic research, in which networks of drug-related behaviors, their underlying pharmacological, physiological, and biochemical mechanism, and particular genomic regions of interest are being identified.[5]

This abstract points to the main topics discussed in the article, rather than providing the specific findings and conclusions of a particular scientific study. For articles that report original experimental work, descriptive abstracts are used rarely. Besides their use for review articles, descriptive abstracts are common for articles that are mathematical or theoretical. Descriptive abstracts are also written for the benefit of those who may be interested in simply retrieving the article rather than in getting the information from it, such as librarians, bibliographers, and scientists searching the literature.

INFORMATIVE ABSTRACTS

Unlike descriptive abstracts, the function of informative abstracts is to report the details of the research and not just to describe what the document contains. Their content focuses directly on the objectives, methods, results, and key conclusions of the research. The following informative abstract describes a graduate thesis. The abstract appeared as the third page of front matter in the thesis, just after the title page and committee signatures page.

Ex. 3.8

Title: A Study of the Butanediols as an Approach to Understanding the Relationship of Alcohol Tolerance to Alcohol Preference in Inbred Strains of Mice

Author: Robert Charles Goldbort

The hypothesis was tested that a positive relationship exists between tolerance to and preference for 1,3-butanediol (1,3-BD) in the high-ethanol-preferring C57BL/6j and low-ethanol-preferring DBA/2j mouse strains, while strain differences in the activity of liver alcohol dehydrogenase (ADH) are small. The C57BL mice showed a significantly higher ($p < .005$) preference for and a greater tolerance to both an excitatory (.0025 ml/gm) and a depressive (.0045 ml/gm) dose of 1,3-BD than the DBA mice. In assays for liver ADH using 1,3-BD and ethanol as substrates, liver extracts from the C57BL strain showed higher specific activities than the DBA extracts for both alcohols, while extracts from the DBA strain dehydrogenated both alcohols at nearly the same rate.

Indiana University of Pennsylvania
May 1975[6]

Though lacking a concluding sentence that weighs the relative influence of neural tolerance and liver metabolism in alcohol drinking behavior, this is a typical informative abstract. A hypothesis is stated, the experimental outcomes are presented, and the reader can see whether the data support or reject the hypothesis. When the thesis research described here was presented subsequently at a scientific conference, the informative abstract was adapted to read as follows.

Ex. 3.9

2838 PHARMACOLOGY

SELECTION OF BUTANEDIOLS BY INBRED MOUSE STRAINS: DIF-
FERENCES IN SPECIFIC ACTIVITY AND CENTRAL NERVOUS SYS-
TEM SENSTIVITY.

Robert Goldbort and R. Hartline (Spon. L. P. McCarty). Indiana University
of Pennsylvania, Indiana, Pa. 15701.

Selection of a 10% (v/v) solution of 1,3-butanediol over water is significantly higher ($P < .005$) among the high-ethanol-selecting C57BL/6j mouse
strain than among the low-ethanol-selecting DBA/2j strain. Measurement of
open-field activity showed the DBA/2j strain to be more depressed after a
high dose (.0045 ml/gm, i.p.) of 1,3-butanediol than the C57BL/6j strain
and to be significantly more active at a low dose (.0025 ml/gm, i.p.) of the
drug. With butanediol as a substrate the specific activity of alcohol dehydrogenase in liver homogenates was greater in the high-selecting C57BL/6j
strain than in the low-selecting DBA/2j strain. These results could account
for the selection and tolerance differences between the two strains if confirmed by *in vivo* analysis.[7]

The description here of the same research for an oral presentation is even more
direct (starting with the results), provides more data and methodological details, and has a more explicit concluding assessment. (Scientific presentations
are described further in Chapter 7.) There is also a significant change in the
language used to describe drinking behavior—from ethanol-preferring to
ethanol-selecting animals—which reflects how scientific usage evolved to
avoid anthropomorphic wording. Finally, this alcohol research was communicated in yet a third way in an informative abstract for a journal article.

Ex. 3.10

GOLDBORT, R., C. W. SCHNEIDER, AND R. A. HARTLINE. *Butanediols: selection, open-field activity and NAD reduction by liver extracts in inbred mouse strains.* PHARMAC. BIOCHEM. BEHAV. 5(3) 263–268,
1976.—Mice from the high-ethanol-preferring C57BL strain and the low-
ethanol-preferring DBA strain were tested for their preference for butanediols. The C57BL strain showed a significantly higher preference for a 10%
(v/v) solution of 1,3-butanediol than the DBA strain. The C57BL strain also
showed a significantly greater consumption of 1,2- and 2,3-butanediol, but

the separation between the strains was smaller than with 1,3-butanediol. Both strains uniformly avoided 1,4-butanediol. Tolerance for 1,3-butanediol was tested in an open-field monitor at 3 doses. At the lowest dose the DBA strain was hyperactive and the C57BL strain was unaffected. At the highest dose both strains were equally depressed. The specific activity of NAD reduction on incubation of liver extracts with 1,3-butanediol and ethanol as substrates was higher with both compounds in extracts from the C57BL strain.

Mice   Butanediols   Tolerance   Preference   NAD reduction   Activity[8]

This third version focuses on comparative pharmacological effects and is broader in scope in that, unlike the earlier versions shown in Exs. 3.8 and 3.9, it includes results obtained with all four isomers of butanediol (1,2-, 1,3-, 2,3-, and 1,4-butanediol). Note also the keywords listed at the end to assist readers in quickly identifying central features of the study. Space limitations permitting, adding an introductory sentence that explicitly states the hypothesis and a concluding sentence on the findings' implications would further enhance this version.

A final example of an informative abstract, for a clinical study published in a nursing journal, shows a sectioned style with boldface headings, ensuring that authors will provide comprehensive and consistent detail.

Ex. 3.11

Objective: To test the contributions of lifestyle and stress to postpartum weight gain after controlling for sociodemographic and reproductive influences.

Design: Longitudinal mail survey with retrospective data on gestational weight gain and prospective data on postpartum weight gain.

Setting: Multicounty community in the midwestern United States.

Participants: After deleting from the sample women who became pregnant again, had confounding medical conditions, or had missing weight data, the sample consisted of 88 predominantly white mothers at 6 months after childbirth and 75 predominantly white mothers at 18 months after childbirth.

Main outcome measures: Weight gain at 6 and 18 months after childbirth.

Results: Maternal race and gestational weight gain accounted for significant amounts of variance in 6-month and 18-month postpartum weight gain. Neither lifestyle nor stress contributed significantly to predicting postpartum weight gains. Gestational weight gain was the most important predictor of postpartum weight gain.

Conclusions: Given the contribution of gestational weight gain to postpartum weight gain, further study is needed of high gestational weight gain.[9]

Another partitioning style used in abstracts is paragraphing without headings. The style and the specific components of an abstract vary across scientific disciplines and with the type of study. Two types of information in the example shown in Ex. 3.11, for instance—"setting" and "participants"—are unique to clinical research. There also is some variance among journals in the abstract's placement and typography.

MEETING THEIR PURPOSE: SUBJECTING ABSTRACTS TO SCRUTINY

Given the limited space typically allotted for an article's abstract, its content must be carefully selected. For the reader's benefit, the abstract must emphasize items of information—techniques, results, or concepts—that are new. Therefore, it is not necessary to include background information or citations, which are already provided in the article itself and waste valuable space that can be used to directly report the research objective and findings in the abstract. When references in the abstract are essential, they should be cited as briefly as possible. Moreover, not only must every piece of information be worth its volume, every single *word* must be scrutinized as well in the process of writing the abstract. Are more words than needed being used to narrate the story of the research? Is the wording ambiguous or imprecise in any way? In reporting findings, is the past tense used consistently? Remember too that journal editors and article reviewers, seeking initial orientation, are likely to first read a paper's abstract and that this first impression, in scientific editor Robert Day's words, "may be perilously close to a final judgment of your manuscript."[10] Especially close attention must be given to the abstract's language and sentences that report the new scientific information. Reviewing the instructions and examples in the journal you intend to submit the article to will help you follow any specific typographical, length, and format features that may be required.

Whatever their type or form, article abstracts serve the interests of readers in the research community in various ways, namely by

- indicating whether the full article would be useful to read;
- being published separately from the article by abstracting services (such as *Biological Abstracts* or *Chemical Abstracts*);

- providing terminology to assist in literature searches by individuals or by literature retrieval specialists for indexes and databases.

For indexing purposes, an abstract may incorporate the full bibliographic citation, as in Ex. 3.10. Since an article's abstract may be read by many people who do not read the article itself, it must be written to stand alone, independently of the article, and still make sense to the reader. The goal in writing the abstract, as Wilkinson puts it, "is to convey as much *new* information as possible to scientists in the same or related discipline in as *few words* as possible—accurately." It takes much effort, patience, and relentless scrutiny of language to write concise, accurate, densely detailed, and readable abstracts. The fact that these qualities are not easy to achieve is affirmed by the reliance of major abstracting services on professional abstractors rather than depending entirely on authors' own abstracts. The writer of an abstract will do well to keep in mind that for most researchers abstracts are their primary source of new information in their discipline. The geologist Scott Montgomery underscores the vital and unique role of abstracts in our electronic age for circulating scientific information: "The abstract is the second most read portion of any paper—and, increasingly throughout science, a crucial publication in its own right. Indeed, abstracts are doubtless *the* most widely exchanged and distributed type of scientific writing in the world today. They are often the only published evidence of conference talks, presentations, and research updates. They are frequently excerpted and republished in reference volumes. They are now included in most online bibliographic databases, a major new aid to research. And abstracts are also forms of 'capital' that scientists trade among themselves almost as readily as they do greetings (or criticisms)." Busy researchers need to know, from reading an abstract, whether they would benefit from going on to read the whole article. Alley quotes Winston Churchill as having said: "Please be good enough to put your conclusions and recommendations on one sheet of paper at the very beginning of your report, so that I can even consider reading it."[11]

THE PRACTICALITY OF WORKPLACE SCIENTIFIC WRITING

The short forms of scientific writing illustrated in this chapter fulfill important workplace demands on a day-to-day basis. Organizations would quickly cease to function coherently and systematically without, for instance, the periodic distribution of internal memoranda at all employee levels. Employees in scientific work settings—whether in education, government, or business— must write such documents with some frequency. Moreover, they are expected to do so in a timely and professional manner. A memorandum that summarizes a researcher's quarterly or annual progress, or a technical inquiry sent to another organization, will be scrutinized by various interested readers. These readers will expect not only a coherent presentation of the practical details that meet the communication's purpose, but also a professional manner of expression. Such workplace writing represents the writer's professionalism and, very importantly, projects an image of that workplace itself. The commitment to effective writing in some corporate settings is seen in requirements that employees, including researchers, participate in technical writing workshops provided by outside consultants. It should be clear that no researcher can afford to take workplace writing as an incidental activity. As we head toward the sole chapter in this book devoted to undergraduate scientific writing, it is noteworthy that some of the types of documents we have covered here also may be taught to upper-level science students, either in their major courses or in scientific writing courses. Science majors receive little or no direct instruction in writing job application materials, an unfortunate reality given that most of them enter the workforce rather than pursuing graduate study. Students can be taught to write reprint requests, inquiry letters, and progress memoranda in the course of their research. If it is true that the quality of writing in the workplace could be better than it is, then we must broaden the writing experience of our undergraduate science majors in ways that will strengthen their preparation for doing science as well as for the daily business of the profession.

UNDERGRADUATE REPORTS IN THE SCIENCES

WHAT IS A COLLEGE REPORT IN THE SCIENCES?

Imagine for a moment researching and writing a report for a college genetics class on the subject of gene translocation in stem cells, versus one for a gothic literature class, say, "Is Victor Frankenstein a Responsible Scientist?" Certain differences in purpose, content, sources, and prose style—objective versus subjective—are likely to spring intuitively to mind. Our society teaches us almost subliminally about the differences in thinking and modus operandi between the cultures of the humanities and science long before college, even as soon as we begin watching television as children. Think of the many images in fiction or film or television commercials of the romantic or sensitive poetic type, versus the mad scientist or simply the systematic researcher coolly and efficiently performing laboratory procedures and collecting data. Such images in public media often seem like caricatures of reality, but they nonetheless point up real professional differences. When we see poets writing and when we see scientists writing we imagine them engaged in radically different kinds of activity. We carry such mental pictures of these professional differences into our formal education. Therefore, it should not be surprising that what is expected in writing a college report in biology versus one in literature are rooted in field-specific processes of inquiry. English majors and biology majors indeed do learn, read, write, and apply their discipline's knowledge in considerably different ways. Their respective cognitive and writing experiences in producing a report are therefore also very different. To define fully what college scientific reports are one must begin with basic questions about their subject, purposes, audiences, and how they are researched, planned, and written. What makes them so different from research reports in humanities disciplines?

Parenthetically, it is noteworthy that students may write on scientific topics outside of science curricula, such as in introductory science courses during their general education. Scientific papers also are written in technical, professional, and scientific writing courses and programs. Students of any major can derive valuable perspectives from writing reports that examine how their anticipated work life may rely on scientific concepts, values, and advances, especially given the strong encouragement in college today of learning and thinking that is inter-, multi-, cross-, or trans-disciplinary. A paper on Frankenstein as a scientist, for instance, could make powerful analogies to the scientific values that have engendered revolutionary advances in our time. Conversely, a report on gene translocation in stem cells could comment on how fictional scientists like Frankenstein suggest standards for responsible scientific conduct.[1]

Whatever the disciplinary context in which one may write a scientific report, the standards of accuracy and truth for scientific information still apply. In essence, a college research report in the sciences is a highly organized, professionally worded, and documented communication of scientific knowledge derived from bibliographic sources. Note that the emphasis is on bibliographic rather than experimental sources of information. Science majors who pursue a laboratory career or a graduate scientific education will see this bibliographic emphasis increasingly shift toward original experimental research as the primary source of information for reports. Some students may afford themselves an early glimpse of this shift through laboratory experiences associated with independent study and internships. Depending on the course or situation, a report's readers may range in expertise from the instructor and classmates to other faculty members and research supervisors.

FEATURES OF SCIENTIFIC REPORTS SHARED WITH OTHER DISCIPLINES

Writing scientific papers involves some basic practices and features that are shared across disciplines and others that are based on the unique values and expectations of the scientific community of researchers. Several elements are common to college research reports in general. First, college research reports typically require bibliographic searches (mostly online these days) that will yield both an appropriately narrowed topic and the published sources to rely upon and cite. Second, it is broadly expected that the report will be organized into key traditional parts: an introduction that explains the topic and thesis; middle parts that present, discuss, and cite the information and ideas; and a concluding section that discusses the upshot or implications of the reported information. Third, research papers across disciplines weave a scholarly tapestry with purposes and presentation modes that describe, explain, argue, and ultimately attempt to narrate and support a scholarly story, for which readers are judge and jury. Fourth, all research reports entail aspects of both process and product. When finished, the document has the layout and look prescribed by academic convention. It appears seamless, not revealing what went before: the dynamic and lived experience of making it, with all the associated decisions. Finally, both the process and the product must have been completed responsibly and ethically. These features of research reports that are shared across the curriculum are part of higher education's emphasis on critical thinking. In addition to the various universal academic expectations, however, a writer of a paper in the sciences (say, for a senior seminar in environmental toxicology) must also follow the ways of the tribe. The conventions and standards for producing a research paper—how the writer derives, thinks about, and conveys the researched information formally—take their own form in the scientific disciplines.

UNIQUE FEATURES OF SCIENTIFIC REPORTS

What, then, are features that are unique to scientific reports? To begin with, the empirical nature of scientific inquiry itself dictates how or when something is viewed as a fact. Scientific facts and concepts have qualities that differ radically from, say, literary or theological facts. Writers of scientific reports must be mindful of the importance that scientific inquiry places on deriving knowledge from observation and manipulation of the physical and natural world. A report on cloning biotechnology for a genetics class, for instance, will risk authorial credibility and lose scientific truth value if a discussion of the ethical responsibilities cites religious dogma. Scientific reports must reflect the fundamental process of doing science—they must be objective and accurate, and must rely on facts, ideas, and thinking that are consonant with scientific professionalism.

This sense of scientific truth value and professional boundaries determines various aspects of scientific report writing as both a process and a product: How a scientific report's topic is delineated, how its thesis is formed and supported, how its information is organized and presented, and how language is used to communicate its content are all shaped by how science works. There are always philosophical challenges in making absolute distinctions among different professional or academic discourses, but the various types of discourse do each have their own special ways of speaking. One framework for understanding the different expectations between scientific writing and literary or creative writing is to list the many particular components of any written work—purpose, scope, audience, voice, use of language, style, and so on—and look at how these two basic approaches differ in their treatment of those details (Table 4.1).[2] While such distinctions are sometimes difficult to maintain, they nonetheless serve to highlight the basic sense that scientific and creative documents ultimately have sharply different intentions. Still, it may be argued that some of these features simply are expressed differently across academic cultures. Disciplines may each, for instance, have their own sense of accuracy and clarity with regard to content, style, and aims. Or, consider the opposition between "composing" and simply "writing"—pointing to a concern with style and expression in opposition to a straightforward telling of facts: First, scientist-writers do express their own individual style, which does require close composition or crafting. Second, while some parts of a scientific report may be relatively straightforward to write, such as experimental methods and results that are thoroughly familiar to the author, careful construction is required in those other parts that interpret, evaluate, and connect to current theory. Because the writer must argue for viewpoints and conclusions *convincingly,* a process of composing—down to the level of words and phrases within sentences—actually is vital. Other distinctions are more readily evident, such as subjectivity versus objectivity of content and language, degree of authorial presence (passive versus active wording), sources of information, and relative use of certain organizational and graphical features.

While the more general or cross-disciplinary expectations for college research reports are rooted in the teaching of the Greek and Roman philosophers of two millennia ago, such as Aristotle and Quintilian, the kinds of distinctions made in Table 4.1 are rooted, philosophically speaking, in the early seven-

Table 4.1 Characteristic differences between scientific and creative writing

Characteristic	Creative Writing	Scientific Writing
Purpose	Expression, exposition	Communication
Generality	Typically more general rather than highly specific and detailed	Highly specific, concrete, detailed, rather than general and abstract, except for theory
Writer vs. subject	Personal, subjective, or objective	Impersonal, objective, object-oriented
Audience	Public or nonspecialized readers	Scientific peers
Rhetorical setting	Writer-centered	Writer marginalized in favor of readers
Form	Intrinsic, author-selected, or shaped during composing process	Extrinsic; determined by convention, material, structure of discipline
Content realism	Reflective, imaginative, imaginary	Observational, factual, reportorial
Form vs. content	Shaped by aesthetic objectives	Constrained by scientific content and purpose
Reader interest	Designed to interest	Inherent in content; readers self-selected
Accuracy and clarity	Not central requirements	Central requirements
Language	Expressive, connotative, vivid	Precise, denotative, concise, plain
Stylistic variation	Used for expressiveness, interest	Avoided, conflicts with precision or clarity
Jargon	Undesirable, except aesthetics	Essential for precision among peers
Passive voice	Proscribed because weak, not direct	Used to focus on object of discourse

(continued)

Table 4.1 *(continued)*

Characteristic	Creative Writing	Scientific Writing
Coherence	Effected by topic sentences and transitional elements	Effected by internal hook-and-eye connections and transitional elements
Process of writing	Largely composing	Typically more straight-forward
Source of material	Writer's knowledge and experience	Discrete body of data and concepts
Graphics	Exceptional, embellishing	Required for empirical demonstration
Format	Integrated; headings not common	Headings important, numerous

teenth century. Followers of Bacon's new scientific philosophy championed his empirically rigorous and mathematically plain style of scientific writing, which focused not on the writer or on the words themselves but on things, on objective and measurable reality. Scientific research reports are a form of writing that is basic to the success and progress of science itself. The strict authorial expectations and responsibilities that they demand reflect the values of modern science as a profession. This important point is underscored in a guidebook published by the National Academy of Sciences (NAS), *On Being a Scientist: Responsible Conduct in Research,* which includes sections titled "Experimental Techniques and the Treatment of Data," "Publication and Openness," and "Authorial Practices."[3] Writing scientific reports encompasses elements of both process and product. The particular nature of the process (what you do) and the product (what you make and submit to an instructor) will vary with the educational situation that sets those parameters. Writers of scientific reports must also recognize the human dimension of the process.

SCIENTIFIC REPORT WRITING AS A HUMAN PROCESS

Scientific information does not communicate itself, nor do automatons communicate it; rather, it inescapably bears the impress of the individual human being(s) responsible for its communication. In short, its successes and its shortcomings are also those of being human. As such, the process of writing a scientific report is neither mechanical nor linear. Sound thinking, a good imagination, and much patience are requisite. While having a sense of stages in the writing process is helpful in keeping one's bearings, the overall process is not so much linear as it is *recursive*. This means that it is necessary to go back and forth among all the various parts of preparing the report, from research to proofreading and everything else in between. Watson noted in his autobiographical account of the elucidation of DNA's structure that "science seldom proceeds in the straightforward logical manner imagined by outsiders. Instead, its steps forward (and sometimes backward) are often very human events in which personalities and cultural traditions play major roles."[4] This is equally true of scientific writing. The writing process is different for every writer and every situation, just as personal writing styles differ even within scientific writing conventions. Writing a college scientific report involves a fluid human process that will yield a product that holds up to the rigorous standards of scientific inquiry. Many public essays and autobiographical accounts by scientists are available to show that scientists are people, too, both in their work and in their writing. Whether a report is written individually or collaboratively, writers must be prepared to grapple with the research and writing aspects of the project as a human effort, not merely as a mechanical gathering of information or making of a product. It is well to expect some setbacks or surprises, changes of direction, and a personal search for both scientific truth and a scientific voice. Readers of a report will not know, as the writer does, the nature of the personal investment and journey that led to the document that lies seamlessly before them.

THE WRITING SITUATION: WHAT IS EXPECTED?

Suppose an instructor gives a class an assignment that requires writing roughly ten pages on just about any topic that relates to the subject of the course, a common scenario. For some students, this can be a nightmarish moment, while for others it may be just an interesting opportunity. Either way, from that moment on the pressures of bibliographic and topic decisions, as well as of constructing the report itself to communicate the information to be found, all begin to weigh on the writer. Where does one begin? Before jumping right into the process of making decisions about the report's topic and other aspects, full awareness is needed of the project's context and situation. In short, what exactly are the expectations? Various situational factors must be understood from the start, including the following:

- What is the report's purpose, scope, length, timeline, and intended audience?
- Should the report be mainly either explanatory or argumentative, focused more on reviewing and explaining scientific theory or on supporting a particular view?
- How does the report relate to the overall content and goals of the class?
- Is collaborative writing required? Peer-critiquing?
- Will it be written for "other" audiences? (Is it part of an internship requirement? Is it cross-disciplinary?)
- What computerized writing environment is available? (This may include computer labs, word-processing and graphics software, printers, scanners, and online research resources through a library or the instructor's Web site.)
- Are there any restrictions on the types of bibliographic sources to be used?
- What criteria will the instructor use to evaluate the report?

As to the question of audience, for instance, the paper's focus and overall value as a writing experience are likely to benefit by conceiving an audience other than just its evaluator (the instructor). A report on complications associated with human birthing—postpartum mood disorders, say, or the effects of smoking—naturally can be directed to women in their childbearing years rather than to the scientific community or a lay readership generally. A clear understanding of the parameters of the writing situation and of the instructor's specific expectations will permit the report-writing process and its stages to proceed more smoothly and confidently.

WORKING WITH OTHERS: COLLABORATIVE SCIENTIFIC REPORTS

Collaborative research writing in the sciences and technology is common. Whether in academic or industrial settings, scientists and other technical researchers routinely conduct their work and prepare reports in team situations. Writing a scientific report collaboratively poses challenges that require special attention, including differences among team members in specific talents and abilities, motivation, commitment, timeliness, and personality. If your project is collaborative, here are some key suggestions for optimizing the team's effectiveness in working together.

- Use a detailed outline to determine the project's scope and to distribute specific responsibilities among the team members.
- Determine the strengths of each team member, such as graphics or editing for style and grammar, and ensure that duties are distributed equitably and fairly.
- Prepare a document template that shows agreed-upon styles and formats for such aspects as multilevel headings, figures, and documentation.
- Create a firm schedule for team members to complete and share their progress on their assigned duties, using focused team meetings and e-mail distribution lists.
- Develop methods, such as distributing minutes of meetings, for keeping close track of which project tasks have been completed and which still need action.
- Take into account each team member's personality in assigning roles and for team activities such as meetings or interviews.

In writing situations, personality and emotional differences can wreak havoc if not anticipated and monitored, or they can be used constructively to the team's advantage. To the extent possible, the team should identify members who would serve best in such roles as initiator, energizer, follower, diagnostician, opinion giver, coordinator, orienter-summarizer, evaluator-critic, procedure developer, graphics designer, and secretary (taking the minutes, for example).[5] Assigning roles based on personality insights will not guarantee a project's overall success, but together with mapped-out procedures it will go a long way toward minimizing potentially disruptive surprises that could slow the project's momentum or jeopardize the final report's quality.

THE RECURSIVE STAGES OF WRITING A RESEARCH REPORT

Because it is a nonlinear process, like scientific inquiry itself as described by Watson, writing a research paper will have you moving back and forth among the various aspects of the research and writing process. This recursive cycle involves the following components, with whatever personal mosaic any given writer may make of them.

- Selecting a topic
- Searching the literature
- Planning and drafting content
- Designing and laying out document-specific features
- Reviewing, editing, revising, and proofreading
- Documenting information

As we consider each of these components, it will help to keep in mind two basic points: First, the writing process is whole and living rather than segmented into separate and discrete parts. Despite any intellectual or practical discussion or artificial division of its elements, it is a seamless, organic, and lived human experience that is irreducible to a tidy set of steps or parts or tools laid side by side, as if for some surgical procedure. Second, writing a scientific report is an adventurous process in the personal sense of being engaged in creating an individual mosaic of choices. The writer cannot fully anticipate the nuances of the individual research and writing experience that lies ahead. It is important to keep this point firmly in mind from the start, even as you begin to weigh alternatives and make decisions regarding prospective topics and sources. Whether your research and writing is yours alone or team-structured, as you embark on this decision-making process, you will need to frame working answers to some basic questions.

RESEARCH AND WRITING: ASKING THE RIGHT QUESTIONS

The process of writing a research paper must begin with an inquiring mind that is open but thoughtful, vigilant, and prudent. The writer must test options and prospective avenues, avoiding dead ends and open waters alike, intent on carving out a supportable scientific narrative of appropriate and workable proportion. As your ideas about a topic and a specific thesis emerge—even as words are set to paper (or to a computer screen)—the process must be guided by a certain sense of direction. Just as scientific inquiry begins with experimental questions, so the process of writing a scientific report must begin with critical thinking that provides answers to some basic questions:

- What are the topic, thesis, and conclusion?
- What points, reasoning, and methodologies support the conclusion?
- How strongly does the evidence support the conclusion?
- How carefully does the language avoid ambiguity?

- How appropriate and helpful for the reader is the demeanor and "voice"?
- Are there any conflicting or questionable assumptions and "facts"?
- What are the limitations of the research and information presented?
- Is the conclusion warranted, given these limitations?
- Is the research and writing process being conducted responsibly and ethically?[6]

Note that this list includes questions about the report's language. How we use language is at the core of how we think, write, read, and process information. Communication of scientific information depends on what the learner does with language. The questions posed above will permit a writer to fully evaluate the focus, thoroughness, and objectivity of the research and writing process judiciously and with dispatch. Sufficient time spent on these initial critical judgments will save much trouble with potential pitfalls later. Any college writing project is defined by its instructional context, by the specific requirements of the course, the instructor, and the assignment itself, so it is necessary to grasp fully the given writing situation.

GETTING STARTED: TOPIC AND SOURCE DECISIONS

Given the typical situation, in which an instructor provides neither a focused topic nor the specific sources to be used, students are left to come up with them on their own, relying on their resourcefulness, imagination, personal interests, and any bibliographic constraints. This is unlikely to be a straightforward process, and there is no single or magical method for choosing a "good" topic. Fortunately, using the many online databases available today will permit a quick initial test of topic ideas for availability of sources. Online searches will also allow efficient refinement and focus of a topic idea. This is not to say that finding the best or "right" sources is any easier than deciding which topic or focus is best, but only that electronic searching, together with both patience and careful management of preliminary time, will make topic decisions go more smoothly. The pressure of course timelines can turn the vast ocean of topics and sources and personal interests into an immobilizing quagmire, but only one thoughtfully selected drop from that open sea of choices will suffice.

Focusing on a topic goes hand in hand with the bibliographic search process. The kind of search process needed will be determined largely by the instructor's leeway regarding allowable sources, whether they are restricted to the current periodical literature or even just journal articles, versus public magazine articles. There may also be course restrictions regarding electronic sources, such as specific types of Web sites or documents, the authority or expertise of which must be assessed. Before starting to search for sources, the instructor's bibliographic guidelines and restrictions must be understood. Once that is clear, the diversity of specialized online databases available on a campus network, together with any special links that an instructor may provide, will open up plenty of research avenues.

Early in the search for topics and sources, the student will gain more control over the process by using online databases to assess the published work that is available. The topic selected, the thesis, the points and evidence offered in support of that thesis, and the conclusions that can be made are all as good as the sources one chooses to consult and cite. An instructor's constraints on types of sources may also limit topic options. Say, for instance, the movie *Jurassic Park* has inspired you to consider writing about chaos theory, which is based on fractal geometry. Or, that media coverage of the controversy over reproductive technology suggests the topic of experimentation with human cloning or stem cells. Or, perhaps current military technology is of interest. Such topics, each for its own reasons, may present source availability problems in the scholarly literature: chaos theory may still be a relatively young field, human cloning is legally restricted, and most information about military technology is likely to be classified.

The opposite problem is common when there is an abundance of information on a topic and an instructor limits types of sources only mildly (such as regarding certain types of Web sites). Many scientific topics, such as those in health care biotechnology, readily yield rigorous scientific sources. Given such availability, it is critical that source types be distinguished as to their type, audience, purpose, and overall scientific rigor. There can be a dizzying and potentially frustrating array of choices: beyond the millions of articles available in journals worldwide, many other types of articles, books, and specialized forms of scientific information and documents—from patents to corporate reports—may turn up on any given topic. Today we have not only the printed documents in traditional libraries, but also specialized databases and the vast amount of material on the Internet. The great diversity in form and purpose of publications that communicate scientific information requires making careful bibliographic distinctions and assessments in the search process.

TYPES OF SOURCES AND THEIR USES

Whether one seeks only a particular type of source, such as current journal articles, or has few such restrictions at all, it is necessary to distinguish the different types of published sources of scientific information. How is a book written by a scholar for a university press different from a book written by a journalist for a commercial publisher, for instance? How is an article on environmental toxins published in the *Journal of Environmental Health* different from an article on "sick-building syndrome" written for a popular magazine like *Time*? Or, what about an article on anabolic steroids in *Joe Wieder's Muscle and Fitness* magazine for bodybuilders versus one in the *American Journal of Sports Medicine*? These are critical distinctions to make because a report will be only as authoritative as the selected sources. Books, articles, and other publications vary widely in purpose, audience, author expertise, level of formality, technical rigor, and documentation (Table 4.2).

There are two important differences between lay sources, such as public magazine articles by journalists, and scholarly sources such as journal articles, and these differences will greatly affect the reliability of any report using them. First, journal articles are primary rather than secondary sources: they are written by the original researchers who generated the knowledge. Second, they are reviewed by peer researchers prior to publication. While it is true that journalistic sources—secondary media removed from the original documents—are easier to read and comprehend, professional scientific sources are more rigorous, dependable, and therefore generally preferable. Before proceeding with our discussion of the research and topic-narrowing process, it will be helpful to take a closer look at the different types of sources that offer scientific information.

SCIENTIFIC INFORMATION IN BOOKS

Scientific information is published in books of different types and by authors whose expertise and authority are not equal to one another. Books that communicate scientific information are written by scientists, journalists, and even by private citizens. This range of authors means that there are differences not only in writers' levels of expertise but also in their books' purpose and audience, comprehensibility and appeal, and reliability and authority as judged by readers. Books are not just books, any more than articles are just articles. In

Table 4.2 Types of sources containing scientific information

Source Type	General Features
Scholarly books	Primarily for professional readers. Includes state-of-the-field critical reviews, historical and theoretical expositions, and original research monographs. Formality, technical rigor, and documentation at highest levels. Peer reviewed.
Textbooks	Used for teaching at all levels. Formality, technical rigor, and documentation vary with student and professional level.
Guidebooks	Multipurpose, including pedagogy and personal use for professional and public readers. Focused on readers' practical needs. Formality, technical rigor, and documentation vary widely.
Reference books	Multipurpose, for professional and public readers. Includes manuals, handbooks, dictionaries, and encyclopedias. Formality, technical rigor, and documentation vary.
Public books	For nonspecialized readers. Includes explanations, personal arguments, histories, biographies, and autobiographies. Formality and technical rigor typically mild; documentation varies.
Journal articles	Communicate current research and ideas. Technical rigor, formality, and documentation at highest levels. Peer reviewed.
Trade magazine articles	Communicate current information and research. Formality, technical rigor, and documentation vary, but typically at high levels.
Public magazine articles	Coverage ranges from broad to special-interest subjects. Formality and technical rigor mild; documentation variable and informal.

(continued)

Table 4.2 (*continued*)

Source Type	General Features
Newspaper articles	Coverage typically broad. Formality and technical rigor mild; documentation variable and informal.
Newsletters	Cover special-interest subjects in short forms for public, professional, and internal workplace audiences. Formality, technical rigor, and documentation mild.
Pamphlets and brochures	Provide practical information in brief forms for public and professional readers. Includes product or service details and special announcements. Formality, technical rigor, and documentation low.
Government documents	Serve public and professional readers. Includes books, reports, and collected statistical data. Formality, technical rigor, and documentation vary widely. Sometimes peer reviewed.
Private-sector publications	Serve public and professional readers. Includes research and development updates, special agency reports, and commercial catalogues. Formality, technical rigor, and documentation vary.

choosing a particular book as a source for a scientific report, one must give careful attention to its intended use. A report on obesity, for instance, will benefit from relying more heavily on a book written by a physician or a scientist who specializes in that field than on one written by a lay author who wishes to share personal experiences and strategies in battling the condition. On the other hand, a journalist's account of testimonials by individuals who have lived with obesity might be used in the report's introduction to draw attention to the personal impact of the problem in everyday life.

Books Published by Scholarly Presses

Another important point about books is that publishers differ just as authors do. The most prestigious, authoritative, and reliable books tend to be those published by university presses (such as the University of Virginia Press or University of Chicago Press). A major reason for this is that such books are "refereed," which means that each manuscript is reviewed and evaluated by experts on the book's subject before it is published. This is also true of books published by professional organizations (such as the American Association for the Advancement of Science or the National Academy of Sciences). Books accepted for publication by university and professional association presses typically represent cutting-edge research and thinking in a field. Any number of examples can be given here, but two books by biologists that are important contributions to scientific scholarship, both published by Harvard University Press, are Edward O. Wilson's *Sociobiology: The New Synthesis* (which stirred much controversy in 1975) and Ernst Mayr's *The Growth of Biological Thought* from 1982. For undergraduate students, however, one disadvantage of these scholarly books is that they are written for professional peers or advanced students and therefore can be difficult to comprehend for those just embarking on their major. University presses also publish textbooks and guidebooks, which tend to be more accessible for students, but much more often these kinds of books are produced by commercial publishers.

Books Published by Commercial Presses

Commercial publishers, such as St. Martin's or HarperCollins, tend to publish books intended for the general public, but sometimes these risk being less reliable, for the primary concern of these businesses is profitability. Compared with books published by scholarly presses, the decision to publish by commercial presses may rely substantially less on extensive peer reviews, due to cost and public appeal factors. Two examples of such books, which sensationalized their topics for public appeal but became controversial due to their questioned accuracy, are *The Hite Report on Male Sexuality,* by Shere Hite, from 1987, and Richard Preston's 1995 book on an Ebola virus outbreak, *The Hot Zone.* For someone interested in the Ebola virus, a book on the subject written by a microbiologist and published by the Centers for Disease Control and Prevention or by a leading university press, rather than one written by a journalist and published commercially, is likely to be more authoritative and dependable scientifically. However, scientists also may write book-length essays or essay collections, sometimes autobiographical, to share their professional insights and experiences with scientific outsiders. Engaging autobiographical works written by Nobel laureates include the biochemist James D. Watson's *The Double Helix* (1968), microbiologist Salvador E. Luria's *A Slot Machine, A Broken Test Tube* (1984), and physicist Richard P. Feynman's *"Surely You're Joking, Mr. Feynman!"* (1985). Commercial presses also publish books by researchers who offer scientific reflections in a personal and informal style, such as Stephen Jay Gould's *Ever Since Darwin: Reflections in Natural History* (1973), Peter B. Medawar's *Advice to a Young Scientist* (1979), and Evelyn Fox Keller's *Refiguring Life: Metaphors of Twentieth-Century Biology* (1995). Or, a book may aim to share the personal passion that an author feels for a subject, as did *The Life of the Bee,* written by Maurice Maeterlinck in 1901, or Vincent G. Dethier's *To Know a Fly,* from 1962.

In addition to full-length books for the public that offer perspectives based on original research by the authors, commercial presses publish textbooks and other reference works, such as manuals. These may be written by important figures in their field, such as James Watson's *Molecular Biology of the Gene,* from 1965, or Robert Hinde's *Animal Behavior: A Synthesis of Ethology and Comparative Psychology,* from 1966. For students, important and current textbooks can provide a useful starting point in finding topics for reports because they cover key information and ideas in a particular field broadly, and therefore may suggest subject keywords to use for online searches. A final thought to keep in mind about books, whether published commercially or by scholarly presses, is the timeliness of their information. It can take years before a book manuscript finally appears in print, so its information may be outdated by the time it reaches readers. For scientific and technological areas in which progress occurs rapidly—such as pharmaceuticals, genetics, and nanotechnology—cutting-edge developments are best followed in current periodicals.

SCIENTIFIC INFORMATION IN PERIODICALS

The multiple annual issues of periodicals allow for much quicker publication, making their information more current than books. Some periodicals are published every week, such as the magazine *Newsweek* or the journal *Science,* and others appear twice a year. Even daily newspapers cover scientific subjects regularly. Articles that communicate scientific information appear in a wide range of periodicals, from scholarly journals and trade magazines to public magazines and newspapers. It is not uncommon for periodicals, including academic research journals, to come and go in the publishing marketplace, or to change their scope, aim, or name. The *Quarterly Journal of Studies on Alcohol* (published every three months) became the *Journal of Studies on Alcohol* (published twice a month), for instance, and *The Technical Writing Teacher* (published three times a year) is now *Technical Communication Quarterly* and publishes articles on technical writing theory and practice as well as teaching. As with books, all periodicals and all scientific articles are not of equal value in their authoritativeness and reliability.

Journal Articles

The most dependable scientific articles—though at the same time the most challenging to read and comprehend—are those published in scholarly or research journals. (Writing a journal article is the focus of Chapter 9.) Journal articles, in particular those that report experimental results versus theoretical or review articles, are distinguishable from other types of articles by various conventional features. These universal features include an abstract following the article's title, citation of colleagues' research, graphical representation of findings, and for most scientific fields the IMRAD structure (introduction, methods, results, and discussion) or variations thereof. It is in academic research journals that scientists typically publish their original experimental work and thereby submit it to review by peer referees. Before a paper is published, its methods, findings, and conclusions are scrutinized for their validity, originality, and value to the scientific community.

Scientists sometimes refer to the body of published research on a subject as "the literature." By this shortened reference they generally mean the original articles written by the researchers themselves and published in their field's professional journals. Academic or research journals are the place for scientists to tell one another about their discoveries, small and large.[7] Researchers also know that any given field has its leading journals, the ones that are considered more reputable than others. Journal articles are indisputably the preeminent source for information on current scientific research; however, two caveats are in order. First, all original scientific research is not necessarily submitted for publication. Given the competitive nature of scientific activity, together with the interplay in our society among science and business and government, there can be a certain level of guardedness and even outright secrecy about some scientific discoveries. When science is done for profit or for national defense purposes, it is not as open as was called for in the founding ideals of Francis Bacon. This is an important practical consideration when choosing a topic for a report. A current topic, such as the health effects of a food substitute like Olestra, may yield less information in the scientific literature than what is actually known because it is guarded by the companies that create and market the product.

The second caveat is that while the unique peer review process for journal articles works quite well, no safeguard works perfectly. Researchers are indeed human, and a small percentage of articles do contain errors or even misleading or fraudulent statements that slip by in the process. The corrective process of scientific inquiry may eventually detect such instances when other researchers cannot duplicate the published findings. One prominent example is an article from 1989 in which Martin Fleischmann and Stanley Pons claimed to have experimentally demonstrated "cold fusion," a sought-after phenomenon having important implications for energy policy and research, but no other researcher has yet been able to duplicate it.[8] The occasional instances of proven fraud serve to remind us that editors, peer reviewers, and readers—being just as human as authors—can be fooled, too (at least initially). As the chemist Carl Djerassi emphasizes in his public writings, scientific progress depends on a shared trust among members of the world's research community.[9] Aside from the immediacy of oral publications like conference papers, journal articles remain the most important source for keeping current in scientific research. The fact that they can be daunting to read should not deter undergraduate students from familiarizing themselves with this key type of source, especially once they have progressed well into their particular major or are considering postgraduate study.

Other Types of Sources in Journals

Besides articles that report original research, scientific journals contain other featured writings that may be cited. These include editorials, commentaries, and perspectives, policy and position statements, news briefs, columns, and letters. Such writings can make important contributions to the openness and proceedings of experimental activity. When perusing a journal's table of contents, note the different types of featured communications. The widely read journal *Science,* for example, published by the American Association for the Advancement of Science, organizes its contents page under the following headings and subheadings.

Ex. 4.1
Departments
    Science Online
    This Week in *Science*
    Editorial
    Editor's Choice
    Contact *Science*
    Netwatch
    New Products
    Inside AAAS
    Science Careers
News of the Week
News Focus
Letters
Books *et al.*
Policy Forum
Perspectives
Technical Science Abstracts
Brevia
Research Article
Reports[10]

Most of the contributions are authored, with the only anonymous ones being the news pieces and some departments (such as the editorial, and new products). Not listed on the contents page, however, is a substantial section in the back pages headed "*Science* Personnel Placement," which lists teaching, research, and administrative openings for scientists in academic institutions, private companies, and government.

Although journals are of primary importance, other types of periodicals have their own unique value as sources for undergraduate scientific reports. Some kinds of information may be less readily available in journals than in periodicals published with different subgroups of readers in mind, such as trade, special interest, and general public audiences. Articles on scientific subjects tailored for such audiences can serve as a useful complement to the information found in journals.

Trade Magazine Articles

Next to journals, a useful source of scientific and technical information is the trade magazine. A major function of trade periodicals is to provide updates on new and important ideas, technologies, and practices in a particular trade or profession. These periodicals typically publish articles by professional experts in a field and therefore are generally very reliable, though the peer-review process may be either weaker or altogether absent, compared with journals. The articles typically are not based on original scientific studies but rely more on information from the author's professional experience and knowledge, companies that develop and market improved technologies and procedures, government documents, interviews, surveys, and questionnaires. Some examples of trade magazines are *Occupational Health and Safety, Plant Engineering, Aviation Week & Space Technology, Corrections Today, Chemical and Engineering News, Metalworking Fluid Magazine, Laboratory Equipment,* and *Patient Care.* While articles in journals and trade magazines are written for academic and professional audiences, other types of periodicals publish articles on scientific subjects for various sectors of the public. These include special-interest magazines and newsletters that are focused topically, as well as broad-based magazines that include articles on scientific subjects of current interest to the public at large. These periodicals vary widely in their depth of coverage, authoritativeness, technical rigor, and reliability, depending on such factors as the publication's purpose, its established prestige, and the level of expertise of the authors.

Newsletter Articles

There is also a wide range of newsletters available. These periodicals are published, in print or online, by commercial presses as well as under such auspices as medical foundations, scientific associations, and university research groups. They are also published by some scholarly or research associations for their own members, like the Society for Literature and Science's *Decodings,* or *News and Notes* published by the American Association for the Advancement of Science. Newsletters publish short items with minimal detail on current research and serve primarily as referential starting points—they may give researchers' names and a short summary of their findings, for example—which can be used for more comprehensive research in the professional literature. The commercially published *Back Letter* offers advice and current news based on information found elsewhere (such as medical conferences) for those who suffer from back ailments. Given the difficulty of ascertaining the reliability of such secondary sources, the information found in a publication like this could be complemented and corroborated by a search for journal articles on musculoskeletal disorders. On the other hand, one need not worry about the reliability of the *Mayo Clinic Health Letter,* published by the Mayo Foundation for Medical Education and Research, or a newsletter published by the Centers for Disease Control and Prevention (CDC), or by the Medical Geography Specialty Group at Penn State University. Many newsletters, such as the highly technical *Medical Science Monitor,* identify each of their publications by both dates and volume and issue numbers, as do journals and magazines.

Special-Interest Magazine Articles

Special-interest magazines range widely in comprehensiveness, rigor, and prestige. For instance, two popular special-interest magazines that publish rigorous articles, often written by professionals in their field rather than by science journalists, are *Scientific American* and *Psychology Today.* These publications may almost be viewed as "soft" journals. Their information is detailed and challenging, but professional jargon and citation of researchers' work is kept at a minimum. However, even the less rigorous special-interest magazines may be useful for research reports. Consider the topic of anabolic steroids. A magazine like *Joe Wieder's Muscle and Fitness,* aimed at bodybuilders, may contain practical information not available in periodicals like the *American Journal of Sports Medicine* or the *Journal of Athletic Training.* Its articles (and advertisements) may contain examples of various types of performance-enhancing substances that are legally available and descriptions of how bodybuilders incorporate them into their workouts. As a complement to this information, studies reported in the journal literature can be used for experimental evidence regarding medical side effects of such substances. Other examples of special-interest magazines are *Vegetarian Times, Popular Science, Popular Mechanics, Diabetes Forecast Magazine, National Geographic, American Forests, Astronomy, Field and Stream, Flying, Wired,* and *Byte.*

Broad-Interest Magazine and Newspaper Articles

For research papers in some undergraduate courses, especially during the first two years of college, instructors may also permit students to use articles on scientific subjects from broad-based periodicals like *Newsweek* and *Time* as well as from daily newspapers. An article on sick-building syndrome from *Time* will likely offer the human-interest side of a subject—how people are personally affected—versus a scientific survey or a laboratory study reported in the *Journal of Environmental Health.* As with other journalistic publications, however, the information in such periodicals typically is secondhand and removed from original sources. Scientific information from such sources (say, on environmental and health effects of toxic dumps or spills) must be corroborated and complemented by primary scientific documents. Few magazine or newspaper articles are comprehensive enough for the detail and precision expected in college research reports.

Private-Sector Articles

Scientific or technical information can also be found in magazines or newsletters published in the corporate sector. Such periodicals may be issued, for instance, by the computer industry, auto manufacturers, airlines, and pharmaceutical companies. While their aim may be in part to provide updated information on various technological or scientific developments, the prime motive of any business entity is financial gain. Therefore, information derived from such publications must be carefully qualified for what it typically is, primarily public relations and product appeal literature. We have all learned the great risks of taking such information at face value, prominent examples being the played-down or masked risks of birth control and tobacco products, and more recently the hazards of dietary substitutes or supplements. Corporate publications may best be used merely as examples of what company representatives have to say, but not as substitutes for scientific documents that are peer-reviewed and published by independent (nonprofit) experts. Two examples of corporate research publications are *The Pfizer Journal* and *IBM Systems Journal.*

ELECTRONIC SOURCES

Having discussed the various kinds of articles that may be used for college research papers, we may now ask: How does one find all these articles? Given the high cost of periodical subscriptions, libraries have a limited selection of titles. Online resources allow broader access and more pinpointed searches for articles. Electronic documents, like print ones, must be carefully assessed for their value and authority. Besides the multitude of personal home pages, sites exist for government agencies, businesses, colleges, public organizations, and professional associations. Specialized databases provide access to such items as biochemical structures (e.g., BioInfo Bank), patents (US Patent and Trademark Office), and medical articles (Medline). Multiple databases containing articles can be searched through a gateway like ProQuest.

SEARCHING FOR SCIENTIFIC ARTICLES

There are many search gateways and databases available through college networks and on the World Wide Web, either cost-free or by subscription, that contain abstracts and full texts of scientific articles. Users may perform selective searches on most of these commercial products not only by subject, but also by periodical title, author(s), publication year, and periodical type (such as public media versus scholarly journals). Some examples of multisubject databases available on college networks are:

- EBSCOhost
- Findarticles.com
- LexisNexis Academic
- OCLC FirstSearch
- ProQuest Direct
- Emerald Fulltext
- ArticleFirst

There are also various bibliographic databases for scientific information, such as GeneralScience Index, AccessScience, Medline, and PsycInfo. Some databases also provide articles as PDF files (which retain the appearance they had in the original publication), a format that is more readable and instructive to students becoming familiar with journal-style articles. The PDF copies of articles in back issues sometimes are available without charge at journal Web sites. Optimal use of electronic resources requires organized and focused keyword searching to determine whether particular topics yield a sufficient and appropriate pool of sources. Besides the standard search boxes, databases may provide further options for narrowing searches.

ProQuest, for example, allows users to do both basic and advanced searches. The basic search screen of ProQuest provides pull-down menus and checkboxes to select databases, date ranges, and types of results (Figure 4.1). Below these settings, an expanded search area (not shown here) offers the additional options of searching by article type (e.g., editorial, review, instructional, interview) and publication type (journals, trade publications, maga-

Advanced Search

Tools: <u>Search Tips</u> <u>Browse Topics</u>

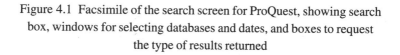

alcohol		Citation and abstract ⬍
AND ⬍	health	Citation and abstract ⬍
AND ⬍		Citation and abstract ⬍

<u>Add a row</u> | Remove a row

Search Clear

Database:

> Multiple databases...
> Business - ABI/INFORM Global
> Education - Education Journals
> Interdisciplinary - Dissertation and Theses
> Interdisciplinary - Ethnic Newswatch (ENW)
> Interdisciplinary - Ethnic Newswatch: A History
> Interdisciplinary - GenderWatch (GW)
> Interdisciplinary - Research Library
> News - National Newspaper Abstracts (3)
> News - The Historical New York Times
> News - The Historical Wall Street Journal
> Science - ProQuest Science Journals
> Social Sciences - Criminal Justice Periodicals
> Social Sciences - ProQuest Social Science Journals

⬍ <u>Select multiple databases</u>

Date range:

> All dates
> Last 7 days
> Last 30 days
> Last 3 months
> Last 12 months
> On this date...
> Before this date...
> After this date...
> Specify date range...

Limit results to: Full text documents only 📄

Scholarly journals, including peer-reviewed 🔊 <u>About</u>

Figure 4.1 Facsimile of the search screen for ProQuest, showing search
box, windows for selecting databases and dates, and boxes to request
the type of results returned

zines, newspapers). Once a basic topic for a report is decided, one useful way
to focus the subject further is by brainstorming to create a list of possible keywords to enter into the search box. Under the subject of alcohol, for example,
the following keywords yielded the number shown of full-text scholarly articles from the past 12 months:

Ex. 4.2	
Alcohol	3,920
Alcohol and health	994
Alcohol and abuse	955
Alcohol and consumption	601
Alcohol and treatment	552
Alcohol and drinking	400

Alcohol and gender	181
Alcohol and law	161
Alcohol and driving	94
Alcohol and binge drinking	88
Alcohol and teenagers	54
Alcohol and genetics	31[11]

These results from a basic and quick search show that sufficient sources are available for a range of viable subtopic choices associated with alcohol. Depending on the scope of the report, more pinpointed searches may narrow the focus to a niche within the broader areas found in an initial brainstorming list.

ProQuest and other software programs permit more sophisticated searching using Boolean methods for stringing keywords together with basic connectives or operators like "and," "or," "not," and "near." Using ProQuest's advanced search screen to extend the keywords "alcohol and health" to "alcohol and health *and heart disease*" lowers the article count from 994 to 46. Expanding the keywords "alcohol and abuse" to "alcohol and abuse *and divorce*" lowers the count from 955 to 20 articles. Or, changing "alcohol and teenagers" to "alcohol and teenagers *and programs*" reduces the count from 54 to 15 articles. These more pinpointed searches yield three prospective topic choices: how alcohol affects the cardiovascular system; the impact of abusive drinking on families; and programs for helping teenagers with drinking problems. University libraries and online sites provide tutorials for Boolean searching. Decisions about when sufficient searching has been done await the scrutiny of the articles themselves to determine their suitability for developing the report. An added convenience of some databases is the option to e-mail selected articles to oneself for perusal later. In addition to using software like ProQuest for searching article databases, one may explore the Web sites of scientific periodicals. Some of these sites—such as the *Journal of the American Medical Association* or *Journal of Cell Biology*—permit free access to the full PDF text of recent articles (charging only for the current issue), while others may charge only a printing fee as if selling individual reprints.

INFORMATION FROM WEB SITES AT LARGE

Besides periodical Web sites and article databases, the Internet offers a vast array of sites containing all sorts of scientific information. These include academic, professional, government, corporate, public, and private sites, as well as individual home pages. Organizations like the American Medical Association, the Institute of Electrical and Electronics Engineers, the American National Standards Institute, and the National Institutes of Health provide links to fact sheets, frequently asked questions (FAQs), pamphlets, press releases, newsletters, position statements, or to their own research documents, periodicals, and books. Such public sites, as well as those of state and local agencies, have links to special reports, statistical information, or legislative proposals on medical and scientific issues that affect the public (e.g., environmental impact studies of industrial pollution, or mother and infant mortality due to postpartum depression). There are also Web sites that provide such highly specialized resources as molecular, genetic, and anatomical databases. The MathMol Library, for one, contains three-dimensional images of many molecular structures discussed in introductory biology and chemistry courses. Biotechnology companies or private foundations may place online various kinds of scientific information, from press releases to technical reports, about their products or research programs. On personal home pages, individual researchers may disseminate descriptions of their research activity and results.

Before using scientific information from a particular site, one must assess the site's authoritativeness and reliability (and perhaps consult with a course instructor). Here are some basic questions for evaluating a Web site's information:

- What is the site's purpose or motive?
- Who maintains the site—that is, does it provide names of authors or organizations?
- Does the site have a bias or an agenda, aside from simply providing information?
- What is the source of the site's or a particular document's information?
- Is the information scholarly, comprehensive, carefully researched and supported?
- Is the information current? Does the site include publication or "update" dates?

A particular site's main use may be simply to provide a reference point for author names or subject terms that can then be used for more comprehensive searches in databases. Some Web sites, such as those of national associations or government agencies, may be used just for current statistics on disease prevalence or updates to legislation. An extra degree of scrutiny may be needed to find the components of an online document that will allow full assessment of its value or bibliographic identification for a report.

PLANNING AND DRAFTING THE REPORT: ANSWERING THE QUESTIONS

Once sufficient information has been gathered to begin drafting the report, it is time to make decisions about how to present the research findings. Here it is necessary to return to those critical questions that guided the search in the first place. What kind of information was sought? Do the sources address each of the points that were necessary to discuss in the report for supporting its thesis? In planning for how the collected information will be used and ordered, a useful visualization technique is to map out the report using an outline.

OUTLINING CONTENT

Why do an outline? Even without actually writing out a formal plan, a focused and meticulous search is likely to begin producing mental images of a report's point-by-point organization. An outline on paper, however, can serve as a concrete checklist, not unlike a pilot's preflight review of the plane's condition. It helps to double-check that every item is in its proper place. An outline's structure and detail will allow the writer to assess how effectively the information will be conveyed and how convincingly the thesis will be supported for its readers. Is there sufficient or too much information, or data, or examples? Are adjustments needed in the report's scope? The process of preparing an outline, in consultation with an instructor and with classmates (whether in collaborative or peer-critiquing groups), provides the writer a more solid sense of the report's measure of quality and success.

How detailed should an outline be? The straightforward answer is: as detailed as is desirable for assisting the writing of the report's draft. Should it be a *keyword* outline, a *sentence* outline, or some combination? An outline with keywords will later facilitate wording of the report's headings and subheadings. A sentence outline may provide topic sentences for starting paragraphs. An outline's level of detail is a personal decision (unless the report is collaborative) to be determined by the sufficiency of road signs that will guide the drafting of the report. Along with the keywords or sentences, one can include various markers, such as approximations of the length in words of each of its sections, author-year citations to show how sources will be used, or notations indicating placement of visuals. In the end, any outline is a customized blue-print to suit one's own needs in planning a report from title to bibliography. At the same time, it must be a flexible tool. It is best viewed as a guide, a plan yet to be fully tested, so the writer must remain open to practical adjustments in the roadmap.

SAMPLE OUTLINE: GENETICS OF ALCOHOL-DRINKING BEHAVIOR

Suppose one is writing an outline for a ten-page report to be titled "Genetics of Alcohol-Drinking Behavior." Searches for current information have yielded a diverse and reliable set of nine sources: three journal articles, a magazine article, two types of monographs, a government report and pamphlet, and a professional newsletter. These will be used to support the thesis that "inherited physiological traits affect individual alcohol drinking patterns." In essence, the report's aim is to explain this genetic link and to convince its readers that the scientific evidence is compelling. A combined keyword-sentence outline for this topic might look as follows.

Ex. 4.3

Genetics of Alcohol-Drinking Behavior

I. INTRODUCTION (1 p.)

Thesis: Inherited physiological traits affect individual alcohol drinking patterns.

Discuss the extent of alcohol problems in our society and worldwide. What insights does alcohol research offer to address and alleviate these problems?

A. Public concerns: youth, families, work (Shalala, 2002)

B. Alcohol-genetics link (Wechsler, 2002)

II. ALCOHOL RESEARCH METHODS AND DEFINITIONS (3 pp.)

Explain how clinical and experimental approaches differ, and what unique kinds of results each has to offer. How do these approaches complement one another?

A. Clinical versus Experimental Methods

   1. Clinical study of alcohol drinking (NIAAA, 2002)

   2. Animal research on alcohol drinking behavior (Dlugos and Rabin, 2003; Rodan et al., 2002)

   [Table 1: Cross-species studies]

B. Special Features of Alcohol Metabolism (Miles, 2000)

   1. Biochemical pathways

---

2. Pharmacological effects
    a. Bipolar CNS effects [Figure 1: Excitation-depression curve]
    b. Tolerance effects

III. LIVER AND NERVOUS SYSTEM STUDIES WITH RODENTS
(5 pp.)
Identify and explain the different kinds of hypotheses, models of thinking, methods, and tools used in investigating liver and neural roles in alcohol drinking. What are the relative influences and importance of liver
and nervous system factors?

A. Liver physiology and drinking behavior (Murphy et al., 2002)
    1. Liver studies and results: metabolic rates
    2. Genetic association: controlling gene loci
B. Central nervous system physiology and drinking behavior (Miles,
2000)
    1. Neural mechanisms: neurotransmitter binding (GABA, sero-
       tonin)
    [Figure 2: Schematic of neuronal alcohol sensitivity]
    2. Inheritance of metabolic rates

IV. CONCLUSION (1 p.)
How far have researchers come in understanding drinking behavior and
its causes, and what health care initiatives do the research outcomes
suggest?

A. Implications of a genetic link (Stocker, 2002)
B. Screening test for alcoholism? (Thiele, 2002)

V. REFERENCES

Dlugos, Cynthia A., and Rabin, Richard A. (2003). Ethanol Effects on
Three strains of Zebrafish: Model System for Genetic Investigations.
Pharmacology, Biochemistry and Behavior, 74(2), 471–480. [journal
article]

Miles, M. F. (2000). Understanding Adaptive Central Nervous System Responses to Ethanol by Use of Transcriptional Profiling. In *Ethanol and
Intracellular Signaling: From Molecules to Behavior,* ed. by J. B. Hock,
A. S. Gordon, D. Mochly-Rosen, & S. Zakhari. Bethesda, MD: Natl. Institute on Alcohol Abuse and Alcoholism (NIAAA) Research Monograph No. 35, National Institutes of Health (NIH), US Department of
Health and Human Services. [monograph]

Murphy, Brenda C., Chiu, Tillie, Harrison, Michelle, Uddin, Raihan K., and
Singh, Shiva M. (2002). Liver and Brain Specific Gene Expression in
Mouse Strains with Variable Ethanol Preferences Using cDNA Expres-

sion Arrays. *Biochemical Genetics,* 40(11–12), 395–410. [journal article]

National Institute on Alcohol Abuse and Alcoholism (NIAAA; Rev. 2002). *Alcohol: What You Don't Know Can Harm You.* Bethesda, MD: NIH, US Department of Health and Human Services. NIH Pub. No. 99-4323. Retrieved September 12, 2004, from http://www.niaaa.nih.gov/. [pamphlet]

Rodan, Aylin R., Kiger, Jr., John A., and Heberlein, Ulrike (2002). Functional Dissection of Neuroanatomical Loci Regulating Ethanol Sensitivity in *Drosophila. Journal of Neuroscience,* 22(21), 9490–9501. [journal article]

Shalala, Donna E. (2000). *Tenth Special Report to the US Congress on Alcohol and Health.* Washington, DC: US Department of Health and Human Services. [government report]

Stocker, Steven (2002). Finding the Future Alcoholic. *The Futurist,* 36(3), 42–46. [public magazine article]

Thiele, Todd (2002). Psychologist Probes the Genetic Secrets of Uncontrollable Drinking. *Center Line,* 13(1), 1. Chapel Hill: Bowles Center for Alcohol Studies Newsletter, University of North Carolina School of Medicine. Retrieved September 15, 2004, from http://www.med.unc.edu/alcohol/cenline/.

Wechsler, Henry (2002). *Binge Drinking on America's College Campuses: Findings from the Harvard School of Public Health College Alcohol Study.* Boston: Harvard School of Public Health. [monograph]

The outline's list of references is close to APA style, but it uses authors' full names, both volume and issue numbers for articles, and original capitalization for titles. When course requirements for citation are flexible, a complete format in an outline will make it adaptable to any style later. An outline having the degree of detail shown here also will constitute a strong test of a topic idea and its thesis. To have come this far, all one is likely to have needed—besides of course critical thinking—is a seat, a notepad (at least to keep track of keywords), a connected computer to search the Internet for full-text sources, a printer, and perhaps a couple of visits to a library to check for current books and current issues of journals. As one begins drafting the report and filling in its sections, a detailed outline will facilitate decisions regarding deletion, addition, or rearrangement of content. More important, it will give the writer more confidence that the questions posed at the project's start can now be addressed successfully and convincingly.

MAKING CHOICES ABOUT REPORTORIAL MODES OF DEVELOPMENT

The outline for the alcohol report represents one option for its development, an analytical approach. A different thesis or focus within that same topic of behavioral genetics, using the same nine sources, could result in a very different structure. Or the same information could be organized using some eclectic or creative approach that draws upon the sources differently. This once again is the personal dimension that that makes the writing process so uniquely dynamic. The writer must decide on the best-suited and most effective option for developing a report with its particular topic and purpose. A scientific report can be developed using one or more of the following methods: inductive, deductive, sequential, comparative, and analytical.

DEDUCTIVE VERSUS INDUCTIVE DEVELOPMENT

Developing a report inductively parallels the sense of movement (not necessarily so linear) in scientific research from articulating a hypothesis about something unknown (a "problem") to experimentally testing it and then generalizing from the results. Analogously, the report writer selects a topic (problem), fashions a thesis, tests it bibliographically, evaluates the research findings, and concludes with inferences or generalizations supported by those results. This inductive method of development is commonly practiced when researchers report their experimental activity in journal articles using the IMRAD model (described further in Chapter 9). Writers who develop a report inductively must realize, Wilkinson says, that "readers do not know the destination until they arrive at it; therefore, they cannot recognize or verify a wrong turn along the way."[12] An inductive narrative must provide a clear, logical, stepwise roadmap so that readers are not forced to retrace their steps or make inferences from insufficient or ambiguous information. Readers trust and expect to be led along toward a report's conclusions responsibly and smoothly. The alcohol report outlined above would proceed inductively as it reveals evidence for the link between alcohol drinking and inheritance. It must build a scientific case for the hypothesized behavior-genetics link, and then conclude with generalizations or inferences that explain the connection and evaluate its significance, implications, and potential applications (such as diagnosing, treating, or preventing alcohol abuse).

Writing a paper deductively means revealing its destination at the start, thereby giving readers a reference point for visualizing and evaluating the path to that conclusion. As the narrative develops, readers can readily see where they agree or pause to check if they are still on track or whether the writer is proceeding along a logical route. Deductive development is well suited for long and complex discussions of concepts and theories. For instance, if the alcohol paper focused on explaining how some gene is hypothesized to influence drinking behavior, readers can assess the logic and evidence offered to show how the gene exerts its behavioral effects. Since deductive development contrasts sharply with the inductive approach that parallels the experimental process, and begins instead with the conclusion, "the writer must be wary of giving the development the authenticity or validity of definitiveness, generality, or universality."[13]

SEQUENTIAL DEVELOPMENT

Some topics are well suited to sequential development, such as those involving temporal events (describing procedures and processes, for instance) or spatial series (smallest to largest objects). Describing three-dimensional entities like equipment (such as a microscope) does not lend itself to sequential development unless the description is based on accompanying visuals (having two dimensions). Then the paper's development can move sequentially from, say, upper to lower parts of the visual (photo), from side to side (block diagram), or from top to bottom (organizational chart).

COMPARATIVE DEVELOPMENT

With a topic that is developed using a comparison structure, the writer compares either a linear series of entities (X to Y to Z) or two or more entities relative to a series of attributes. A simple comparative approach with the alcohol report would be comparing three rodent species (mice to rats to hamsters) relative to a single attribute, for example their inherited neural sensitivity to ethanol. The paper could then proceed simply to discuss that attribute sequentially in each species. An example of a more elaborate approach is comparing the neural effects of three different alcohols (ethanol, propanol, butanediol) in the three species. Such a comparison could be developed either vertically or horizontally. In a comparison that is structured vertically, each species might be discussed individually and sequentially (mice followed by rats and then hamsters) relative to the neural effects of all three alcohols. In a horizontally structured comparison, all three species are taken together to compare the neural effects sequentially with each alcohol. In any case, comparative reports are readily adapted to a linear process of scientific exposition.

ANALYTICAL DEVELOPMENT

The alcohol report also could be developed as an analysis of some basic scientific problem or concept. For instance, how would a gene operate to affect a particular animal behavior like drinking a pharmacological agent such as alcohol? Topics that are developed analytically may require consideration of a web of complex relationships, such as those associated with chronology, logical explication, cause and effect, and comparisons. Such a web of interrelationships can be treated like a two- or three-dimensional object, represented diagrammatically, and adapted to linear exposition.

Whatever the report's method of development, the writer is likely to engage in some combination of four basic compositional modes, namely, description (actions, objects), explanation (theories, logic), argument (alternative viewpoints), and narration (events, natural phenomena). Along the way, the writer will bring to bear facts, data, ideas, examples, logic, and personal ingenuity to convince readers of the validity of the report's thesis, evidence, and conclusions. The methods chosen to develop the report and to achieve its purpose, together with the writer's command of technical language, will determine how readers respond to it as a scientific document.

BEGINNINGS, MIDDLES, AND ENDINGS

The simple and sensible idea that a written (or oral) communication must have a beginning, a middle, and an end is found two millennia ago in Aristotle, and it applies in specialized ways to scientific reports. Tradition dictates that the standard parts of a report are the introduction, the sections that comprise its body of findings from the scientific literature, and a concluding discussion.

INTRODUCTION AND BACKGROUND

The introduction to a scientific report sets up the reader's expectations by providing a blueprint for what its writer intends to accomplish. An effective orientation to the subject's significance and to the aims of the report should do the following:

- identify a topic's scope—that is, what it will describe, explain, argue, or narrate;
- provide context for the topic's significance, namely, an overview of relevant research, theory, and practice;
- raise key ideas, concepts, or terminology applicable to the subject;
- state the report's proposition or thesis;
- delineate the specific objectives (subpoints) in support of the thesis.

Beyond delineating the topic's scope and providing background on its scientific significance, the writer must explain clearly the report's purpose. In this regard, the report's objectives must be differentiated from its thesis. In the following example, the first sentence states objectives and the second a thesis— a position or claim that the writer must convincingly demonstrate to be sustainable.

Ex. 4.4

1. This report will compare two competing theories, hepatic versus neural bases, which researchers use to account for genetic differences in the alcohol drinking behavior of laboratory mice.
2. The comparison will be used to suggest a middle ground not sufficiently tested, namely: Each animal's neural sensitivity to alcohol works in concert with its own rate of liver detoxification to shape drinking patterns with a "synergistic" effect that is like a behavioral fingerprint.

The objectives are what the writer intends to do or cover in the report; the thesis tells why the information is significant. In papers that are experimentally derived, a thesis is replaced by the concept of a testable hypothesis. A complete and clear introduction provides a smooth transition to the report's middle sections, which present the bibliographical findings to develop and support the writer's key message.

SECTIONS ON RESEARCH FINDINGS

Even after a plan has been outlined and the report's method of development has been determined, one must remain open to any necessary adjustments in structure or content. In presenting the findings, a recursive process becomes important. Does the plan need to be adjusted? Are more sources needed for better development or support of particular points? In any case, an effective presentation of the findings should:

- take up each point or aspect in some logical and apparent order (e.g., spatial or temporal features, strongest to weakest evidence);
- interconnect key points with one another and with the thesis coherently;
- support points concretely (with appropriate data, examples, cases, logic, analogies);
- reconcile or address opposing viewpoints that are relevant to the thesis;
- tell *and* show by complementing verbal reportage with visual representation;
- partition major aspects liberally but judiciously using headings and subheadings;
- recognize informational limitations (the writer's, the report's, the readers').

A clear and plain presentation will be reflected in how authoritative and convincing readers perceive the narrative to be. This is particularly important when a point is being addressed on which experts offer conflicting data or interpretations. In the alcohol report, for instance, studies may be cited that support the role of either liver or nervous system biochemistry as a key inherited influence on drinking behavior. Although such opposing positions may indeed be difficult to reconcile, the writer can assist readers to this end by including visuals—representing experimental data or theoretical models—that clarify the divergent reasoning of each camp. Scientific facts and ideas that are communicated logically, visually, and readably will permit readers to comprehend more readily how the report's conclusions are grounded in the documented sources.

DISCUSSION AND CONCLUSIONS

The concluding section of a research report is not a mere formality. It is an opportunity to pull together the scientific narrative with a sense of closure regarding the significance and implications of the research findings. A scientific report's ending should do some combination of the following:

- reaffirm the thesis, with its scientific significance (practical, theoretical);
- underscore the major scientific points covered in support of the thesis;

- reach conclusions from scientific evidence that validates or modifies the thesis;
- offer recommendations implied by the conclusions (e.g., on workplace practice, public policy, legislation, legal issues, ethics, or further study).

A report's closing should leave readers with a convincing demonstration of the rigor and authority of its information as well as of the validity of its conclusions. In the alcohol report, the writer may reach a conclusion that inheritance is less important than social factors in influencing drinking behavior, notwithstanding the value of animal studies. Therefore, based on the sources that present clinical studies, the report's conclusion may recommend comparative study of psychosocial factors that affect drinking in different cultures or subgroups in societies.

While an academic research report in any subject typically must have appropriate content in its beginning, middle, and ending sections, the expectations for scientific reports are unique and highly formalized. Certain types of statements and information are expected to be in their traditional parts of the research report, in a manner that parallels experimental thinking. Scientific conventions also apply to bibliographic documentation (Chapter 5) and to the incorporation of visual matter (Chapter 6). A scientific report also tends to be highly segmented by extensive use of headings throughout its text.

PARCELING A REPORT'S CONTENTS WITH HEADINGS

Besides a report's internal dynamics for developing a topic and for writing effective prose, there is an external technique for guiding readers through content: dividing and subdividing information using a system of headings and subheadings. In partitioning a topic and identifying its key elements, headings are superimposed signposts. Readers are helped by headings in various ways, such as in finding parts of the paper that may be of special interest to them or when a paper's development is unconventional or especially complex. Because headings stand outside the scientific narrative itself, their removal should make no difference to the text's meaning, though their absence will make readers expend more energy following and decoding the narrative. The detailed and four-tiered outline (I, A, 1, a) for the alcohol report facilitates a multilevel sectioning system. The outline's second major (or primary-level) section, for instance, allows the following heading structure:

Ex. 4.5
ALCOHOL RESEARCH METHODS AND DEFINITIONS
Clinical Versus Experimental Methods
    Clinical study of alcohol drinking
    Animal research on alcohol drinking behavior
Special Features of Alcohol Metabolism
    Biochemical pathways
    Physiological effects
        Bipolar CNS effects
        Tolerance effects

The style of the headings must allow readers to recognize each division level. In the following example from a journal article, the methodology section (shown with partial text) is divided using a three-tiered hierarchy of headings.

Ex. 4.6

METHOD

Animals

A total of 300 male mice, half from the C57BL/6J strain and half from the DBA/2J strain, were obtained from the Jackson Laboratory, Bar Harbor, Maine. All animals were 10–12 weeks old at the time of testing.

Chemicals

All chemicals were obtained from commercial sources and were used without prior examination for purity; 1,3-butanediol and 1,4-butanediol, Eastman-Kodak Company; 2,3-butanediol, J. T. Baker Chemical Company; 1,2-butanediol, a gift from Dow Chemical Company.

Procedure

Preference testing. Preference testing was carried out in a windowless room with the light cycle and temperature held constant. Sixty naïve mice from each strain were tested with 10% (v/v) solutions of the four alcohols (15 mice from each strain per alcohol).

Activity tests. Activity was monitored in an open-field apparatus previously described [6]. Animals were tested for 15 min exactly 30 min after an IP injection of 1,3-butanediol or saline, and monitoring began immediately upon introduction of the animal into the open field.

Preparation of the liver homogenate. Five animals from each strain were sacrificed by cervical dislocation and the livers removed, weighed, and ho-

---

mogenized in 9 volumes of cold 0.25 M sucrose for 2 min at 5° C with a Potter-Elevhjem homogenizer.

Assay of NAD reduction. Assaying crude extracts of liver for dehydrogenase activity with a substrate such as 1,3-butanediol and its presumed immediate metabolic oxidation product β-hydroxybutyraldehyde poses two problems that make attempts to determine individual dehydrogenase activities no more informative than evaluating the ability of the extracts to reduce NAD with the alcohol as the substrate.[14]

Note that each level of heading uses a different style to distinguish it from the others. Typography, positioning, and spacing are typical features used for multilevel text segmentation. Heading systems may also use numbers to distinguish the various levels, or differences in type size. Although numerical and letter-size systems are used less frequently in college reports, they can be especially helpful for navigating other types of documents, such as lengthy government reports. Headings for scientific reports do not use color, unlike such technical documents as procedural or equipment manuals (such as red headings for sections that caution about special hazards or troubleshooting certain problems).

Two other important aspects of headings are their wording and their relation to a document's text. Wording in headings typically is compressed ("Preparation of liver homogenate"), or just a single word may suffice ("Chemicals"). It is generally inadvisable to use full sentences or questions as headings. And because the headings are meant to serve only as guideposts for the reader, they are considered *outside* the narrative, not a part of it. Therefore, the sentence of any particular section does not follow directly from the words in its heading, but rather from the preceding section's last sentence. So if a heading reads "Preparation of the liver homogenate," for instance, the first sentence following it ought not begin "*It* was prepared by . . ." but rather "*The liver homogenate* was prepared by . . ."—as if the heading did not exist. Despite their extratextual nature, well-designed headings do provide structural guidance for moving through a document's contents efficiently or selectively and will earn the writer gratitude for saving the reader time and energy.

ADDITIONAL ELEMENTS FOR REPORTS THAT ARE FORMAL

In some instructional situations, especially in technical or scientific writing courses, students may be asked to prepare research reports that are of the formal type. Formal reports contain features that are typical in corporate, government, and institutional settings. While the text or body of these reports have the traditional content and structure already described, there are other parts used to increase their formality. The more evident features are covers and binding, but the primary additional elements that make a report formal are referred to as the front and back matter.

FRONT MATTER

Elements preceding a formal report's body provide information that orients its intended readers about the report's purpose, context, and content. The front matter also is a chance not only to impress upon readers the significance of the information itself but to show the care and professionalism with which the document was prepared—something especially important in the business or administrative side of the scientific work world. The typical components of front matter are:

- *Front cover:* title, author(s), date, and graphical elements
- *Title page:* title, author, and author affiliation
- *Transmittal memorandum:* explanation of report's aim, scope, and content
- *Table of contents:* list of all sections and subsections, with their page numbers
- *List of visuals:* list of all tables and figures with their number, title, and page
- *Abstract or executive summary:* encapsulation of the report's content and conclusions

The design and layout of each of these features vary considerably within the expected conventions. For instance, cover graphics may be absent or range from being conservative or subtle to dominant and colorful, though they should always be tasteful, inoffensive, ethical, and culturally sensitive. There may be legal considerations involving appropriate use of licensed logos. Practices also vary regarding the abstract or executive summary, in both length and detail. The transmittal memo, typically a single page that explains the report's impetus, purpose, and contents, may also include relatively informal or editorial commentary and special acknowledgements.

Here is a transmittal memo addressed to a university dean that might be included in a formal report by a committee charged with special duties.

Ex. 4.7
Fillmore University
College of Arts and Sciences

To: Dr. Janice N. Trudeau, Dean, College of Arts and Sciences
From: Dr. Samuel E. Hillary, Chair, CAS Program Review Task Force
Date: December 5, 2002
Re: Final Report and Recommendations

In fulfilling your charges when you appointed the Program Review Task
Force on January 12, 2002, we hereby submit our final report. You asked us
to "study how we can reorganize the College's scientific programs to meet
21st century curricular needs." This report describes how we proceeded to
meet that charge and offers our recommendations.

During the past eleven months, the Task Force met 18 times. We assessed
the specific need to reorganize and expand our scientific programs, especially in the Life Sciences. The Committee distributed a survey among faculty and students for their input. This valuable feedback guided our discussions and helped us develop a working plan and timetable for achieving the
changes delineated in our final report.

It was our pleasure to serve you and the CAS faculty, and we hope that our
findings and recommendations will help lead our science programs in the
necessary direction.

More than a minor formality, the transmittal memo introduces a document into
the administrative archive officially, so that further action can be taken. The
memo also provides the gist of the report for busy readers who must prioritize
their day-to-day work activities. The same professional attention must be
given to the typical items that are placed in the report's back matter.

BACK MATTER

Following the report's main text—with its findings, conclusions, and recommendations—several kinds of items can be included as back matter, such
as the following:

- *List of references or bibliography:* lists all sources cited or consulted, sometimes accompanied by brief annotations
- *Glossary:* lists and defines technical terms used in the report

- *Supplementary graphics:* tables and figures that further illuminate the text
- *Mathematical information:* formulas, special derivations, and statistical data
- *Sample documents:* brochures, Web sites, FAQs, and fact sheets

Back-matter items that follow a list of sources typically are titled as appendixes, for example, "Appendix A: Glossary." The format and sequence of back-matter items vary. Glossaries may read across the page or in columns, with the term on the left and its definition in the right column. A list of references may precede or follow appendixes, which may have their own citations. The selection and style of back-matter items must be guided by the particular needs of the report's audience(s).

WRITING THE DRAFT AND MEETING READER EXPECTATIONS

The draft stage naturally is the point of thorough testing of a topic. Following a detailed outline like the one in Ex. 4.3, as well as remaining mindful of the report's readers, will go a long way toward keeping the writer on track. The writer should assess the draft as it develops by asking basic questions concerning the purposes and methods discussed in this chapter, for each component of the report; examples of such questions are listed in Table 4.3. Depending on the specific writing situation, additional questions may arise as the draft is continually reviewed (with instructor or peer critiques). Here we return to our beginning: the characterization of a college research report in the sciences as a highly structured, professionally worded, and documented transmission of scientific information. Readers of a scientific report expect that it will present the information coherently, with a clear interrelationship among its parts, and that the writer will follow the strictures of scientific English proficiently and responsibly. Gauging a paper's overall success requires asking the right questions about its fundamental parts and traditional features. In assessing the draft the writer must also examine particularly its grammar, usage, and readability. The key criterion by which scientists judge the value of language is not its capacity for expressiveness in subjective or literary senses, but rather its practical utility. Once a draft is completed, guided by all the factual, structural, and linguistic expectations, the remaining task is to polish it into a final copy by double-checking it, standing, as it were, in the shoes of both writer and reader.

Table 4.3 Questions for critiquing a report at the draft stage

Report Aspect	Questions for Assessing Draft
Introduction	• Are the topic's focus and current significance explained fully? • Is there background information on current scientific activity and thesis? • Is the thesis statement appropriately focused and clearly stated? • Are the points to be covered in support of the thesis delineated?
Findings	• Is each objective or point sufficiently, clearly, and rigorously covered? • Is the content presented in an accessible and thought-provoking manner? • Is there sufficient scientific support (examples, data) for each point?
Discussion and conclusions	• Are the thesis and key scientific findings reaffirmed? • Does the conclusion point up the findings' practical or theoretical implications? • Are the limitations and remaining questions regarding the findings assessed? • Does the report end thought-provokingly (e.g., by looking to the future)?
References	• Are the sources authoritative, unbiased, mainly primary? • Are expectations being met for source types (e.g., current journal articles)? • Are any Internet sources carefully screened (or even pre-approved)? • Is documentation precise and stylistically consistent?
Visuals	• Are the visuals selected, designed, and incorporated carefully? • Are visuals fully labeled with a number, title, caption, and credit? • Does any visual need a legend to identify symbols or colors?
Format	• Are multilevel headings used logically to segment information? • Are heading titles and subtitles brief, informative, and parallel? • Are the report's layout, design, and typography appropriate and effective?

(continued)

Table 4.3 (*continued*)

Report Aspect	Questions for Assessing Draft
Readability	• Is technical jargon minimized and glossed sufficiently? • Is there awkward or ambiguous wording? • Is wording simple, clear, concise, direct, concrete, objective? • Are there coherence devices, such as emphasis of key points and transitions?
Grammar and usage	• Are there spelling, punctuation, typographic, or other mechanical errors? • Are scientific terms and numbers used properly and written correctly? • Are verb tenses used accurately (e.g., in findings versus conclusions)? • Is active versus passive wording used where appropriate? • Is wording concrete and denotative versus abstract and connotative? • Is language biased or inappropriate (e.g., gender, culture, ethnicity)?

FINAL COPY: REVIEWING, EDITING, REVISING, AND PROOFREADING

A final double-checking of the report means evaluating its overall readability, from its content, organization, and language to its use of visuals and typography. Again, the questions used in assessing the draft provide the broad strokes for a starting point in the editing process. The actual work of editing your report can make use of electronic resources while also applying human judgment.

COMPUTERS AS EDITORS

The computer is almost taken for granted as a tool for editing. It is a convenient and efficient tool for electronic cutting and pasting as well as using language aids like a thesaurus and spelling, grammar, and style checkers. For scientific writing in particular, specialized functions and software provide discipline-specific features for creating or editing mathematical formulas, chemical and anatomical structures, or engineering drawings. Separate software packages are commercially available that contain scientific dictionaries, spellers, bibliographic stylers, and proofreaders. Examples of such software, also used for documents more professionally advanced than college reports, are Inductel Scientific and Technical Dictionary, SciProof, and Scientific WorkPlace.

Computer-aided writing does have its limitations. Software for checking spelling, grammar, and style is not error-free, foolproof, or comprehensive. A spell-checker will flag typos—words with missing letters or inverted letters, unfamiliar letter combinations or words, or double-word errors like "the the"—to help in correcting these efficiently and quickly. On the other hand, it will not flag a missing word like an article, as in "drank [the] fluid," or a correctly spelled word that is used incorrectly or poorly chosen ("two," "too," and "to," or "effect" and "affect"). A spell-checker will not catch an inadvertent use of "phase" for "phage" or "animal infections" in place of "annual inspections," errors that surely will confuse if not chill readers. Grammar and style checkers are helpful in finding such items as unbalanced marks (quotes, parentheses), use of the passive voice, or single-gender referents; however, they also flag unusual or innovative but technically correct wording. As dazzling and convenient as all the ever-improving electronic resources may be, they cannot substitute for the creativity and ingenuity of the human mind. Computers cannot write, either literally or figuratively, for us.

HUMANS AS EDITORS

We return, then, to the proposition that writing is a human experience. In the editing process, writers must face two complicating human realities: The first is that it is not easy to objectify language to the extent required by science, and the second is that the individuality of the writer cannot be surgically removed from scientific writing any more than one can control the individuality of the reader's interpretation of the text. However, the primacy of intention is a fundamental criterion for a scientific report. The reader of a scientific report must decode the specific technical meanings intended in the writer-researcher's exposition. In contrast, the intentions of the literary writer may become irrelevant as the work takes on a life of its own as an art object to be individually experienced. The essential task in editing a scientific research report is to root out ambiguities so that writer and reader can share the same understanding of the text.

The second and related truth scientific writers face is that no amount of editing will eliminate the individual person in any text. It is not simply a matter of removing personal pronouns or feelings. As Luria recognized, there is latitude within science's linguistic strictures. The inherent risks of such latitude, however, require vigilance to ensure that the writer's individuality is not self-pointing. The issue is not whether scientific writing is a human act that is subject to human limitations, but whether the truthfulness, honesty, and professional integrity of the writer, process, and product have been preserved and protected to the extent that is humanly possible. A meticulous draft-editing process goes a long way toward helping a report's writer survive the academic and human challenge of producing a successful scientific report.

BEING THOROUGH FROM START TO FINISH

We close as we began, namely, by underscoring the point that writing a scientific research paper is a recursive human experience having unique professional qualities and expectations. In the process, writers may even experience a surprise or two that alters the course of the research or challenges preconceptions about a subject. As to the quality and effectiveness of the final product, Michael Alley uses a sports metaphor to remind us to keep our eye on the ball at all times: "Finishing a paper is much the same as finishing a baseball game. Some teams, when they're ahead, let up during the last few innings. They play sloppily, sometimes so sloppily that they lose their lead. Some writers are the same way. They work hard on the first few drafts, and then let up on the final drafts, allowing typos to pull down their work."[15] Any let-up in the professional rigor demanded by a thorough editing and proofreading process, causing inattention to even the smallest or seemingly insignificant details in content, format, or language, is likely to be costly down the road. At the very least, doubt may be raised in readers as to the writer's professional standards. Along with the challenges of researching and writing a report, there is also a sense of adventure in not knowing exactly where one will wind up until the report's final copy is submitted. In that sense, the experience of writing science is like that of doing science: controlled and constrained but nonetheless flexible and open to the unexpected. In the vital process of sharing what they do and learn, researchers also collaborate as a community of writers and readers—every scientist is both—to preserve the integrity and effectiveness of their professional communication and thereby of their unique mode of inquiry itself.

DOCUMENTATION OF SCIENTIFIC SOURCES

THE IMPORTANCE OF BIBLIOGRAPHIC DOCUMENTATION

The advancement of scientific inquiry, the very process of experimental research itself, depends on a trusting collaboration among its practitioners. Thoroughness and honesty in citing the scientific literature is an integral part of that professional collaboration. When a research paper is prepared, whether it is a college assignment or a professional article, its information must be connected to the scientific archive in that field by the citation of the relevant published work of fellow researchers. In this way, not only are readers provided with the broader scientific context of the work being reported, but credit is also given for past research and to those who did it. Beyond being an ethical standard professionally, giving credit to words and ideas that originated with others (by using quotation marks, for example) is a legal necessity to avoid committing plagiarism, defined these days as the theft of "intellectual property." The National Academy of Sciences offers this caveat to researchers: "Failure to cite the work of others can give rise to more than hard feelings. Citations are part of the reward system of science. They are connected to funding decisions and to the future careers of researchers. More generally, the misallocation of credit undermines the incentive system for publication."[1] Finally, in addition to being a professional, ethical, and legal responsibility, citation (or lack of it) also reveals the writer's degree of authority on a subject. How well does the writer know the depth and scope of the related work that is already published? What is the relation of the researcher's original contribution to the larger and established body of knowledge in that field?

Once decisions have been made about which sources to cite, there is the matter of which citation style to follow. In most cases, the style already will have been determined, either by a particular instructor or by in-house guidelines followed by a particular periodical. Chemists follow the style manual published by the American Chemical Society (ACS), biologists have available to them the guidelines published by the Council of Biology Editors (CBE), biomedical authors may use the American Medical Association's style guide, and various disciplines (including anthropology) follow the University of Chicago's style manual. In addition, research periodicals provide their own style guidelines (often online), which may differ from those of the various widely used manuals. Besides following the prescribed format with precision, there are basic considerations in the citation process itself.

CITING RESPONSIBLY: SELECTIVITY, ACCURACY, AND COMPLETENESS

Any writer of a scientific paper must use a selection process in citing sources. One may read and know much more than is appropriate or necessary to cite. Only those sources need be cited that are directly relevant and centrally important to the purpose of the research. The writer therefore must establish criteria for limiting references. Which publications contain key findings that are associated directly with the subject of the paper? Being selective will help avoid a distracting series of citations like this:

Ex 5.1

Various factors, such as carbon dioxide emission (2, 5, 18, 25, 38, 43–45), ozone depletion (12, 23, 29, 35–37, 51), and rain forest destruction (4, 9, 15, 33, 41, 51, 62), must be considered in projecting the extent of global warming.

When the citation possibilities are extensive, it is better to say that many researchers have worked on the problem and then to cite some of the most pertinent sources. The criteria for source selection can include originality, importance, comprehensiveness, and balance—the earliest papers, for instance, or seminal works, review articles, and articles that represent cutting-edge work on the subject. Most of the remaining references are expendable, especially if they are cited indirectly in review articles or more specifically elsewhere in the paper. In essence, the necessary citations are those that contributed directly to the report and its findings, including ones that may conflict with its thesis. The various contributions of sources may include concepts, theories, recommendations, statistical data, equations, and experimental measurements.

Beyond selectivity in citing, the bibliographic information that is supplied must be absolutely accurate and complete. Source citations help readers retrieve the references. Avoiding errors and omissions is not only a courtesy but also a professional responsibility that saves much time and energy, considering the collective expense (across the research community and into the future) required in the search for even one cited source. All bibliographic information should be recorded from the source itself, in hand, rather than citing from other bibliographic citations. In our computer age, some researchers find it convenient to use citation software, such as Citation and Endnote. While using such software will save time and effort, the accuracy and completeness of citations will still be a function of how carefully you entered the data. When recording a source's information, misspelling an author's name, mistyping the year, or omitting a series number, subtitle, or edition will simply be reproduced electronically, leaving the errors and giving readers serious difficulties.

Whether citing electronically or manually, only the utmost care will ensure the reliability of bibliographic information, beginning with legible notes from the original source that contain every available piece of information to identify it without question, and followed by equally close transcription into a document.

EXAMPLES OF CITATION STYLES IN A LIST OF REFERENCES

Typically, complete information for any cited source is provided at the end of a paper in a section titled "References" or "Works Cited." "Bibliography" carries a different sense, namely, that there is no obligation to cite the listed sources in the paper and that the writer need not even have read them. A bibliography provides readers with a comprehensive or representative list of sources on a subject. In contrast, a list of references or citations contains all the sources that the writer actually read and cited in the paper. Citing a source carries a greater responsibility for knowing the specific information it contains. It is also permissible to have both types of lists, one for sources cited and another for those consulted, such as may be done in review papers or when a specialized textbook or a field-specific dictionary has provided the writer with general background information. The scientific community as a whole practices a range of citation styles. This chapter illustrates four of the most widely used styles in the sciences, along with a humanities style for comparison, published in the following manuals:

- *Scientific Style and Format,* Council of Biology Editors (CBE)
- *The ACS Style Guide,* American Chemical Society (ACS)
- *The Chicago Manual of Style* (CMS), University of Chicago Press
- *Publication Manual,* American Psychological Association (APA)
- *Handbook for Writers of Research Papers,* Modern Language Association (MLA)[2]

Because all of these manuals are readily available and cover the various citation practices extensively, only a few basic examples will be provided here. Whichever citation style is used in a paper, there are certain items of information that are typically included for any given type of source. The most frequently cited types of sources in a scientific paper are other scientific papers. Less often, writers of scientific papers cite other source types—monographs, textbooks, government documents—that contain special information such as statistics or innovative techniques not available elsewhere. When following any given style, it is necessary to proofread meticulously for these four practices:

- *accuracy:* correct information in all fields, from authors to pagination;
- *completeness:* conventional document identifiers;
- *correctness:* precision in following prescribed guidelines;
- *consistency:* same citation manner for each particular source type.

Whereas students typically receive bibliographic guidance from their instructors, authors of articles must consult a given periodical's guidelines. Since articles are the main types of citations in scientific papers, they will be illustrated first, followed by books and a range of other types of sources.

CITATION STYLES FOR ARTICLES

Scientists publish articles primarily in scholarly journals, which they refer to collectively as the scientific literature. In gathering information for citing an article, attention must be given to the following bibliographic items:

- Author(s)
- Title of article
- Title of periodical
- Year, month, day, season
- Volume number
- Issue number
- Supplement or series number
- Page range
- Electronic information (Internet site, database, date retrieved)

Scientific articles also may be published in collections, requiring additional citation items. All identifiers of the source should be written out completely in one's notes (or software) rather than abbreviated in any way. This will allow adapting the information for any citation style without having to retrieve the source again.

This example shows a basic citation of a journal article with a single author:

Ex. 5.2

CBE   Lamont LS. Dietary protein and the endurance athlete. Int Sports J 2003;7(2):39–45.

ACS   Lamont, L. S. Dietary Protein and the Endurance Athlete. *Int. Sports J.* **2003**, 7 (2), 39–45.

CMS   Lamont, Linda S. 2003. Dietary protein and the endurance athlete. *International Sports Journal* 7, no. 2: 39–45.

APA   Lamont, L. S. (2003). Dietary protein and the endurance athlete. *International Sports Journal, 7*(2), 39–45.

MLA   Lamont, Linda S. "Dietary Protein and the Endurance Athlete." <u>International Sports Journal</u> 7.2 (2003): 39–45.

Note the variations regarding the following stylistic practices: full versus abbreviated author names and journal titles;[3] capitalization in article titles; use of periods, commas, and colons; boldface type (year in ACS), italics (ACS, CMS, APA), or underlining (MLA) in periodical titles; format for volume or issue numbers (the 7 and the 2 in these examples); elision in page ranges; and spacing between items. An article's title, for instance, is capitalized sentence style (only the first word of the title and subtitle) in CMS and APA, while in ACS and MLA it is headline style (each main word capped). Some chemistry journals omit the title because it is not needed for locating the article. (Not shown in any examples here is the first-line indentation for APA and MLA styles.)

This next listing is for a journal article with multiple authors:

Ex. 5.3

CBE Steinberg FM, Bearden MM, Keen CL. Cocoa and chocolate flavonoids: implications for cardiovascular health. J Am Diet Assoc 2003;103(2):215–23.

ACS Steinberg, F. M.; Bearden, M. M.; Keen, C. L. Cocoa and Chocolate Flavonoids: Implications for Cardiovascular Health. *J. Am. Diet. Assoc.* **2003**, 103 (2), 215–223.

CMS Steinberg, Francene M., Monica M. Bearden, and Carl L. Keen. 2003. Cocoa and chocolate flavonoids: Implications for cardiovascular health. *Journal of the American Dietetic Association* 103 (2): 215–23.

APA Steinberg, F. M., Bearden, M. M. & Keen, C. L. (2003). Cocoa and chocolate flavonoids: Implications for cardiovascular health. *Journal of the American Dietetic Association, 103*(2), 215–223.

MLA Steinberg, Francene M., Monica M. Bearden, and Carl L. Keen. "Cocoa and Chocolate Flavonoids: Implications for Cardiovascular Health." Journal of the American Dietetic Association 103.2 (2003): 215–23.

When there are many authors of a work, among the styles illustrated here only ACS and MLA require the inclusion of all author names, no matter how many there are. The other three styles—CBE, CMS, and APA—limit the number of authors listed. APA style is to list up to six authors, followed by "et al." if there are more than six. CMS style lists up to seven, followed by "et al.," and CBE lists up to ten, followed by "and others" if there are more. Note also that at the other end of the spectrum, when the authors are unknown, in four of the styles the citation simply begins with the title, while in CBE the title is preceded by "[Anonymous]."

The following citation is an article from an edited collection of articles that were published previously (1949–1988) in the journal *Science,* and now compiled thematically:

Ex. 5.4

CBE   Gordis L, Gold E. Privacy, confidentiality, and the use of medical records in research. In: Chalk R, editor. Science, technology, and society: emerging relationships. Washington, DC: American Association for the Advancement of Science; 1988. p 143–6.

ACS   Gordis, L.; Gold, E. Privacy, Confidentiality, and the Use of Medical Records in Research. In *Science, Technology, and Society: Emerging Relationships;* Chalk, R., Ed.; American Association for the Advancement of Science: Washington, DC, 1988; pp 143–146.

CMS   Gordis, Leon, and Ellen Gold. 1988. Privacy, confidentiality, and the use of medical records in research. In *Science, Technology, and Society: Emerging Relationships,* edited by Rosemary Chalk. Washington, DC: American Association for the Advancement of Science, 143–146. Originally published in *Science* on Jan. 11, 1980.

APA   Gordis, L., & Gold, E. (1988). Privacy, confidentiality, and the use of medical records in research. In R. Chalk (Ed.), *Science, technology, and society: Emerging relationships* (pp. 143–146). Washington, DC: American Association for the Advancement of Science. (Original work published in *Science* on Jan 11, 1980)

MLA   Gordis, Leon, and Ellen Gold. "Privacy, Confidentiality, and the Use of Medical Records in Research." Science, Technology, and Society: Emerging Relationships. Ed. Rosemary Chalk. Washington, DC: American Association for the Advancement of Science, 1988. 143–6.

Inclusion of the original publication date of the republished article generally is optional, and in any case it is information already provided by the collection's editor.

Citation of a magazine article versus a journal article typically is simpler, not only due to it usually having fewer authors (often just one), but also because volume and issue numbers are dispensable, since magazines typically use only month and day identifiers. However, when volume or issue numbers are given, it is acceptable and desirable to cite magazine and journal articles in a parallel fashion. In the following example of a magazine article, volume and issue numbers generally are not included (except for APA, which requires the volume number).

Ex. 5.5

CBE   Thorne AG, Wolpoff MH. The multiregional evolution of humans.
Scientific American 1992 Apr:76–9, 82–3.

ACS   Thorne, A. G.; Wolpoff, M. H. The Multiregional Evolution of Humans. *Sci. Am.,* Apr 1992, pp 76–79, 82–83.

CMS   Thorne, Alan G., and Milford H. Wolpoff. 1992. The Multiregional Evolution of Humans. *Scientific American,* April, 76.

APA   Thorne, A. G., & Wolpoff, M. H. (1992, April). The multiregional evolution of humans. *Scientific American, 266*(4), 77–79, 82–83.

MLA   Thorne, Alan G., and Milford H. Wolpoff. "The Multiregional Evolution of Humans." Scientific American Apr 1992: 76+.

Given the common practice in magazines of placing intervening material, such as advertisements, with articles, there are different styles for designating discontinuous pagination, from providing the full page ranges (CBE, ACS, APA) to giving just the start page (CMS) or start page with a plus sign (MLA). Whether citing an article from a journal or a magazine, it is desirable to include all standard identifiers—unless proscribed, such as the article's title in some chemistry journals. The importance of articles in scientific communication, whether written for peers or for the public, demands the highest standards of professionalism in following the conventional practices for giving credit, authority, and verifiability for the shared information.

CITATION STYLES FOR BOOKS

Although articles typically make up the largest part of a scientific paper's list of references, researchers also cite various kinds of books. These may include an important textbook, a laboratory technique manual, a seminal work that synthesizes current thought in a field, or a monograph that makes an original contribution in some experimental niche. "Book" also can apply to smaller documents like pamphlets or major reports. For citing a book, one should attend to the following bibliographic items:

- Author(s), editor(s), and translator(s) if applicable
- Title
- Number of edition

- Year of publication
- Publisher name and location
- Pagination (if citing a specific chapter, for instance)
- Volume or series number

For a book published by a committee within a government agency, the gathered citation information should include the committee name, its chair's name, and a publication number. When a book is part of a series, one may note the series title and its general editor(s)—for example, Greg Myers, *Writing Biology: Texts in the Social Construction of Scientific Knowledge,* 1990, published by the University of Wisconsin Press in its Science and Literature Series, with George Levine as general editor. This textbook reference shows the basic elements for citing a book with a single author in a subsequent edition.

Ex. 5.6

CBE   Watson JD. Molecular biology of the gene. 3rd ed. Menlo Park, CA: W. A. Benjamin; 1976. 739 p.

ACS   Watson, J. *Molecular Biology of the Gene,* 3rd ed.; W. A. Benjamin: Menlo Park, CA, 1976.

CMS   Watson, James D. 1976. *Molecular biology of the gene.* 3rd ed. Menlo Park, CA: W. A. Benjamin.

APA   Watson, J. D. (1976). *Molecular biology of the gene* (3rd ed.). Menlo Park, CA: W. A. Benjamin.

MLA   Watson, James D. Molecular Biology of the Gene. 3rd ed. Menlo Park, CA: W. A. Benjamin, 1976.

When the book has editors, the citation begins with their names (except in ACS), as in the following example of an edited monograph published by a government research institute.

Ex. 5.7

CBE   Hoek JB, Gordon AS, Mochly-Rosen D, Zakhari S, editors. Ethanol and intracellular signaling: from molecules to behavior. Bethesda, MD: National Institute on Alcohol Abuse and Alcoholism, National Institutes of Health, US Department of Health and Human Services;

> 2000. NIAAA Research Monograph Nr 35. NIH Publication Nr 00-4579. 210 p.
>
> ACS   *Ethanol and Intracellular Signaling: From Molecules to Behavior;* Hoek J. B., Gordon, A. S., Mochly-Rosen, D., Zakhari, S., Eds.; National Institute on Alcohol Abuse and Alcoholism (NIAAA) Research Monograph No. 35; National Institutes of Health (NIH) Publication No. 00-4579; U.S. Department of Health and Human Services: Bethesda, MD, 2000.
>
> CMS   Hoek, Jan B., Adrienne S. Gordon, Daria Mochly-Rosen, and Sam Zakhari, eds. 2000. *Ethanol and intracellular signaling: From molecules to behavior.* Bethesda, MD: National Institute on Alcohol Abuse and Alcoholism Research Monograph No. 35, National Institutes of Health Pub. No. 00-4579, U.S. Department of Health and Human Services.
>
> APA   Hoek, J. B., Gordon, A. S., Mochly-Rosen, D., & Zakhari, S. (Eds.). (2000). *Ethanol and intracellular signaling: From molecules to behavior.* (NIAAA Research Monograph No. 35, NIH Publication No. 00-4579). Bethesda, MD: National Institute on Alcohol Abuse and Alcoholism, U.S. Department of Health and Human Services.
>
> MLA   Hoek, Jan B., Adrienne S. Gordon, Daria Mochly-Rosen, and Sam Zakhari, eds. <u>Ethanol and Intracellular Signaling: From Molecules to Behavior</u>. Bethesda, MD: National Institute on Alcohol Abuse and Alcoholism, U.S. Department of Health and Human Services, 2000. NIAAA Research Monograph No. 35, NIH Publication No. 00-4579.

An individually contributed chapter in a monograph is cited parallel to Ex. 5.4 for an article in a collection.

SOURCES OTHER THAN ARTICLES AND BOOKS

In addition to articles, books, and parts of books, researchers may cite a range of other types of scientific sources, a few of which are illustrated here. In journals, in addition to the articles there are research letters, news briefs, columns, book reviews, and editorials that can be cited. Other works are dissertations or theses, conference papers, abstracts, and patents. With less frequency, research documents also may cite legislation, legal documents, press releases, and newspapers.

The following are the different styles for citing a dissertation.

Ex. 5.8

CBE   Goldbort RC. 1989. Scientific writing and the college curriculum
      [PhD dissertation]. East Lansing (MI): Michigan State University;
      1989. 204 p.
ACS   Goldbort, R. C. Scientific Writing and the College Curriculum.
      Ph.D. Dissertation, Michigan State University, East Lansing, MI,
      June 1989.
CMS   Goldbort, R. C. 1989. Scientific writing and the college curriculum.
      PhD diss., Michigan State Univ.
APA   Goldbort, R. C. (1989). *Scientific writing and the college curricu-*
      lum. Unpublished doctoral dissertation, Michigan State University,
      East Lansing.
MLA   Goldbort, Robert Charles. "Scientific Writing and the College Cur-
      riculum." Diss. Michigan State U, 1989.

If a dissertation or a master's thesis has been published by a press or by a service like University Microfilms International (UMI), the citation should include the following kinds of additional identifiers: the publisher's name and
location, the UMI number, and the volume and page numbers of Dissertation
Abstracts International (DAI). Prospective readers must know if a dissertation
is available other than in the degree-granting institution's libraries or archives.

Another common type of source that is cited with some frequency in scientific research documents is a conference paper or its abstract, as shown here:

Ex. 5.9

CBE   Goldbort R, Hartline R. Selection of butanediols by inbred mouse
      strains: differences in specific activity and central nervous system
      sensitivity [abstract]. In: Fed Proc 34(3), Abstracts, Federation of
      American Societies for Experimental Biology 59th annual meeting;
      1975 Apr 13–18; Atlantic City (NJ). Bethesda (MD): FASEB; 1975.
      p 720. Abstract nr 2838.
ACS   Goldbort, R.; Hartline, R. Selection of Butanediols by Inbred Mouse
      Strains: Differences in Specific Activity and Central Nervous Sys-
      tem Sensitivity [abstract]. In *Federation Proceedings,* 34(3), Ab-

---

> stracts of the 59th Annual Meeting of the Federation of American
> Societies for Experimental Biology, Atlantic City, NJ, Apr 13–18,
> 1975; FASEB: Bethesda, MD, 1975; Abstract 2838, p 720.
>
> CMS   Goldbort R., and R. Hartline. 1975. Selection of butanediols by in-
> bred mouse strains: Differences in specific activity and central ner-
> vous system sensitivity. Abstract. *Federation Proceedings* 34, no. 3:
> 720.
>
> APA   Goldbort, R., & Hartline, R. (1975). Selection of butanediols by in-
> bred mouse strains: Differences in specific activity and central ner-
> vous system sensitivity [Abstract]. *Federation Proceedings, 34*(3),
> 720.
>
> MLA   Goldbort, Robert, and Richard Hartline. "Selection of Butanediols
> by Inbred Mouse Strains: Differences in Specific Activity and Cen-
> tral Nervous System Sensitivity." Federation Proceedings 34.3
> (1975): 720. Abstract. Item 2838.

The conference paper citation in this example can be lengthy because the abstract was published in a proceedings volume, so that in CBE style it is treated
like a chapter in a book, with publisher name and location. However, the citation also has elements of a journal article citation because *Federation Proceedings* has volume and issue numbers. Another helpful identifier is the abstract number.

Given that experimental research leads to the development of new physical
and biochemical technologies with practical and commercial applications,
scientists also cite patents for these inventions. A patent issued by the United
States Patent and Trademark Office (USPTO, Arlington, VA) is cited as follows:

> Ex 5.10
>
> CBE   Guri AZ, Patel KN, inventors; Plant Cell Technology, Inc., assignee.
> Compositions and methods to prevent microbial contamination in
> plant tissue culture media. US patent 5,750,402. 1998 May 12.
>
> ACS   Guri, A. Z.; Patel, K. N. Compositions and Methods to Prevent Mi-
> crobial Contamination in Plant Tissue Culture Media. U.S. Patent
> 5,750,402, May 12, 1998. Appl. 460703.
>
> CMS   Guri, A. Z., and K. N. Patel. 1995. Compositions and methods to

---

prevent microbial contamination in plant tissue culture media. US
patent 5,750,402, filed June 2, 1995, and issued May 12, 1998.

APA   Guri, A. Z., & Patel, K. N. (1998). *U.S. Patent No. 5,750,402*. Arlington, VA: U.S. Patent and Trademark Office.

MLA   Guri, Assaf Z., and Kishor N. Patel. Compositions and Methods to
Prevent Microbial Contamination in Plant Tissue Culture Media.
Plant Cell Technology, Inc., assignee. Patent 5,750,402. 12 May
1998.

The basic identifying fields required in all the styles are the names of the in-
ventors, the patent number, and the patent's date of issue. Most styles also include the patent's title (except APA). Additional identifiers, as shown above,
may include the filing date (CMS), application number (ACS), assignee
(CBE, MLA), and the patent's official source (APA). If the information is from
the USPTO Web site, the citation includes electronic identifiers.

ELECTRONIC CITATIONS

Many kinds of scientific information can be retrieved electronically today
from a range of locations.[4] These sites include specialized databases and home
pages for government, corporate, educational, and professional organizations.
Retrieving information from some sites, such as scientific periodicals or sensitive government databases, may require either subscription or registration
(with name and password). Beside the basic information shown in the examples above, electronic citations require the name of the database (or Web site),
the uniform resource locator (URL) for the document or search site, and the
date you retrieved or viewed the document.

This first example is a journal article retrieved from a database:

Ex. 5.11

CBE   Jiang R, Manson JE, Stampfer MJ, Simin L, Willett, WC, Hu, FB.
Nut and peanut butter consumption and risk of Type 2 diabetes in
women. JAMA 2002;288(20):2554–60. Available from: http://
proquest.com. Accessed 2003 Oct 2.

ACS   Jiang, R.; Manson, J. E.; Stampfer, M. J.; Simin, L.; Willett, W. C.;
Hu, F. B. Nut and Peanut Butter Consumption and Risk of Type 2

Diabetes in Women. *JAMA* **2002**, 288 (20), 2554–2560. ProQuest database (accessed Oct 3, 2003).

CMS   Jiang, R., J. E. Manson, M. J. Stampfer, L. Simin, W. C. Willett, and F. B. Hu. 2002. Nut and peanut butter consumption and risk of Type 2 diabetes in women. *Journal of the American Medical Association* 288, no. 20 (November 27): 2554–60. http://gateway. proquest.umi.com/openurl?ctx_ver=z39.88-2003 &res_id= xri:pqd&rft_val_fmt=ori:fmt:kev:mtx:journal&genre=article& rft_id=xri:pqd:did=000000249985191&svc_dat=xri:pqil:fmt= html&req_dat=xri:pqil:pq_clntid=954 (accessed October 3, 2003).

APA   Jiang, R., Manson, J. E., Stampfer, M. J., Simin, L., Willet, W. C., & Hu, F. B. (2002). Nut and peanut butter consumption and risk of Type 2 diabetes in women. *Journal of the American Medical Association, 288*(20), 2554–2560. Retrieved October 2, 2003, from Pro-Quest database.

MLA   Jiang, Rui, Joanne E. Manson, Mcir J. Stampfer, Simin Liu, Walter C. Willett, and Frank B. Hu. "Nut and Peanut Butter Consumption and Risk of Type 2 Diabetes in Women." <u>JAMA</u> 288.20 (2002): 2554–2560. *ProQuest.* 2 Oct. 2003 <http://proquest.com>.

Beyond the basic citation, the required electronic identifiers vary somewhat. Only CMS requires the source's full URL, which can be long, as seen above. It is generally sufficient to provide the more limited address of the search site (e.g., http://proquest.umi.com), or just the name of the search product (like ProQuest or Medline). Note that retrieving a copy of an article (such as a PDF image) from a database is different from retrieving an article from an Internet site that publishes the periodical. Citing an article or an abstract directly from an e-journal site does not need database or search product names, although either the article's or the search screen's URL is still included. Abstracts cited from special databases—a search product like ProQuest or a specialized publication like *Chemical Abstracts*—are treated the same as electronic articles but are identified by adding "abstract" in parentheses or brackets after the article's title or with the retrieval date.

Researchers and students today also make use of "aggregated" databases, available through various electronic formats, such as a CD-ROM placed on a university server and accessed through the supplier's Web site. The article cited in Ex. 5.11, for instance, available through ProQuest, actually was retrieved from a "PA Research II Periodicals" database, but this information is virtually ignored since citations do not require it. Given the diversity of electronic sources and the evolving methods of retrieving information from them, the styles for citing them may continue to develop as they become more standardized like those for non-electronic or print sources.

The following example shows citation styles for an entire electronic book, in this case a monograph with a corporate author as well as editors.

Ex. 5.12

CBE   Beers, Mark H; Berkow, Robert, editors. The Merck manual of geriatrics [monograph on the Internet]. 3rd ed. Whitehouse Station (NJ): Merck;2000–2003. Available from: http://www.merck.com/pubs/. Accessed 2003 Nov 18.

ACS   Beers, M. H., Berkow, R., Eds. *The Merck Manual of Geriatrics,* 3rd ed. Merck: Whitehouse Station, New Jersey, 2000–2003. http://www.merck.com/pubs/ (accessed Nov 18 2003).

CMS   Beers, M. H., and R. Berkow, eds. 2000–2003. *The Merck manual of geriatrics,* 3rd ed. Whitehouse Station, NJ: Merck & Co. http://www.merck.com/pubs/ (accessed November 18, 2003).

APA   Beers, M. H., & Berkow, R. (Eds.). (2000–2003). *The Merck manual of geriatrics* (3rd ed.). Whitehouse Station, NJ: Merck & Co. Retrieved November 18, 2003, from http://www.merck.com/pubs/.

MLA   Beers, Mark H., and Robert Berkow, eds. The Merck Manual of Geriatrics, 3rd ed. Whitehouse Station, NJ: Merck & Co., 2000–2003. 18 Nov. 2003 <http://www.merck.com/pubs/>.

The most current print version of the CBE manual does not include the URL in its citation examples for electronic sources, using just the site's name and the retrieval date, but in the example above the URL is given as part of the "availability" information.[5]

In addition to the most frequently cited electronic documents—articles, abstracts, and books (or their parts)—researchers may cite various other types of online sources. These include online reports (governmental, public, corporate, academic), specialized databases, commercial software, organizational and personal home pages, newsgroups, discussion list servers (listservs), and even e-mails (cited as personal rather than published communications, with the date and writer's affiliation). Among the electronic sites more frequently used by researchers are those of government agencies and professional organizations, which often upload research data or statistics as a matter of public information (on population, infant mortality, environmental quality, or scientific policy, for example). Here is a listing for a professional association's home page:

Ex. 5.13

CBE   AWHONN Website [Internet]. Washington, DC: Association of Women's Health, Obstetrics, and Neonatal Nurses; c2002 [cited 2003 Mar 14]. Available from: http://www.awhonn.org/.

ACS   Association of Women's Health, Obstetrics, and Neonatal Nurses Home Page. http://www.awhonn.org/ (accessed Mar 2003).

CMS   Association of Women's Health, Obstetrics, and Neonatal Nurses. http://www.awhonn.org/ (accessed March 14, 2003).

APA   Association of Women's Health, Obstetrics, and Neonatal Nurses Web site. Retrieved March 14, 2003, from http://www.awhonn.org/.

MLA   Association of Women's Health, Obstetrics, and Neonatal Nurses. Home Page. 2002. 14 Mar. 2003 <http://www.awhonn.org/>.

The styles are all simple and straightforward, with only CBE including the association's geographic location. Other items allowable in a home page citation are copyright date (CBE, MLA) and when it was last updated. This final electronic citation is for a database from a government Web site.

Ex. 5.14

CBE   MedlinePlus [Internet]. Bethesda (MD): National Library of Medicine (US); [updated 2001 July; cited 2003 Sep 3]. Available from: http://www.nlm.nih.gov/medlineplus/.

ACS   MedlinePlus database. National Library of Medicine Home Page. *http://www.nlm.nih.gov/medlineplus/* (accessed Sep 2003).

CMS   MedlinePlus. National Library of Medicine. http://www.nlm.nih.gov/medlineplus/ (accessed Sep 3, 2003).

APA   MedlinePlus. United States National Library of Medicine. Retrieved September 3, 2003, from http://www.nlm.nih.gov/medlineplus/.

MLA   MedlinePlus. U.S. National Library of Medicine. Home Page. 12 Sep. 2003. 5 Oct. 2003 <http://www.nlm.nih.gov/medlineplus/>.

Like the association home page cited in Ex. 5.13, a database citation from a government Web site tends to be simple, with the key elements being the database's name, the government agency's name, and update and retrieval dates.

The styles for electronic citation are evolving with the medium. In its 2001 supplement for citing Internet sources, the National Library of Medicine noted that its guidelines were "intended to be evolutionary in nature. As new types of Internet documents are discovered, and as readers submit suggestions, the text will be revised and expanded."[6] Given the many types of Web sites and documents available in cyberspace, researchers will need to exercise patience and ingenuity in keeping pace with these changes for their bibliographic purposes. When there is doubt regarding how to cite any particular electronic source, or about which identifiers to include, one can simply rely on common sense and use whatever Web site or source information is available. Depending on the source to be cited, one will need to gather these kinds of bibliographic elements:

- *URL:* What is the source's unique electronic address on the Internet?
- *Retrieval date:* When was the source retrieved or read?
- *Title or name:* What is title or name of the source, periodical, or Web site?
- *Author/editor/translator/compiler:* Who is responsible for preparing the source?
- *Dates:* When was the source published, copyrighted, or updated?
- *Publisher:* What press, group, or person made the information available?
- *Geographical location:* Where is the publisher, agency, or association located?
- *Type of source:* If not apparent; is it an abstract, a pamphlet, a report, software?
- *Numerical information:* Are there volume, issue, page, or document numbers?

These and any unique identifiers of the source, cited in a conventional style, will permit unambiguous documentation and easy retrieval. Given the constant change characteristic of the Internet and its contents, whenever possible it is advisable to print out the source (or the part cited) to have hard evidence of its existence and what it said.

CITATIONS IN TEXT, IN VISUALS, AND IN BIBLIOGRAPHIC NOTES

End-of-text citations in a works-cited list also require an accompanying style for citing those sources in the text. The basic formats for internal citations are as follows (note that ACS has three options).

Ex. 5.15

CBE   Hamsters are known to be aggressively territorial.[1,4–7,11]
ACS   a. Hamsters are known to be aggressively territorial.[1,4–7,11]
      b. Hamsters are known to be aggressively territorial (*1,4–6*).
      c. Hamsters are known to be aggressively territorial (Hu, 1996;
      Jones and Ulm, 1998; Brown, 2000; Hall et al., 2002).
CMS   Hamsters are known to be aggressively territorial (Hu 1996; Jones
      and Ulm 1998; Brown 2000; Hall et al. 2002).
APA   Hamsters are known to be aggressively territorial (Hu, 1996; Jones
      and Ulm, 1998; Brown, 2000; Hall, et al., 2002).
MLA   Hamsters are known to be aggressively territorial (Hu; Jones and
      Ulm; Brown; Hall et al.).

In numerical styles, the references may be superscripted, parenthetical, or (not shown here) bracketed. Author-year styles differ from one another with regard to various items, such as:

- Punctuation (use of commas and semicolons)
- Number of names listed before "et al." is used (two in ACS; three in CBE, CMS, and MLA; six in APA)
- Multiple works by the same author(s) in the same year (e.g., Griffin 2002a, 2002b)
- Different authors with the same surname (use of first name or initials)
- Anonymous works (use of "anonymous" versus part of the title)
- Page numbers (only when quoting, versus anytime in MLA)

Some styles, including CMS, permit a parenthetical citation to include a brief note—for instance, "(Johnson 2003; only seven subjects tested)." In the author-year system, citations also may be worked into a sentence:

Ex. 5.16
Yamamoto (2003) concludes that "sweetness is discriminated from other tastes by different receptor sites on taste bud cells, a different subset of fibers in the taste nerves, and different projection zones in the brain" (S8).[7]

Sources may also be cited for various purposes in textual locations other than in the text or in a reference list. For instance, a bibliographic note may be placed either at the bottom of a page (as a footnote) or at the end of the text (an endnote). In a "discursive" note, citations may be part of a broader discussion. When the citation in a note (or in a reference list) includes commentary directly about the source, it is called annotated. Footnotes or endnotes (with or without supplementary discussion) generally are not favored in scientific papers, but they sometimes are used. Research articles in the journal *Science,* for instance, include an end-of-text list that serves a dual purpose titled "References and Notes." Here is a listing from an article in *Science* that combines a citation with a note.

Ex. 5.17

32. D. A. Gailey, R. C. Lacaillade, J. C. Hall, *Behav. Genet.* **16**, 375 (1986); that the female's locomotion systematically decreases over the course of courtship was also shown (although not genetically dissected) by T. A. Markow and S. J. Hanson [*Proc. Natl. Acad. Sci. U.S.A.* **78**, 430 (1981)].[8]

The citation style used in *Science,* as in many other scientific periodicals, differs somewhat from the styles we are considering here (such as in punctuation, use of boldface, and omission of the article's title).

Another extratextual location in a research document where a citation may be placed is with a figure or other visual representation. The following are three instances of citations in visuals (two in titles and one in a caption, respectively).

Ex. 5.18

1. Fig. 1. **Age-stratified seroprevalence of mumps antibody during the pre-vaccine era in England and Wales** (*ref. 28*), **Netherlands** (*ref. 29*), **St Lucia** (*ref. 31*), **Poland** (*ref. 33*), **Singapore** (*ref. 30*), and **Saudi Arabia** (*ref. 32*).

2. Figure 2. The "five A's" of smoking intervention. (Adapted from Fiore MC, Bailey WC, Cohen SJ, Dorfman SF, Goldstein MG, Gritz ER, et al. Smoking Cessation. Clinical Practice Guideline No. 18. *Rockville, Md: US Department of Health and Human Services, Public Health Service, Agency for Health Care Policy and Research, Centers for Disease Control and Prevention.* April 1996:22–25. AHCPR Publication No 96–0692.)

3. Fig. 5. Ratios of last to first appearances for brachipod species in the Upper Permian of China. Numbers above bars give total species known

---

> from each interval. The abbreviations L and U signify upper and lower
> substages of the Maokouan (M), Wujiapingian (W), and Changxingian
> (C) intervals [data from (15)].[9]

Style guides generally prefer citations to be placed outside the figure or table, either above or below it—in the caption, in a footnote, or in a source line— rather
than with a figure's legend or in a table's field, to avoid confusion with the visual's content. Although often used interchangeably, the terms "caption" and
"legend" are not synonymous. A caption is a phrase, a full sentence, or several
sentences (which may follow the figure's title) used to explain the figure concisely. It is typeset and placed immediately below the figure. A legend (or key)
identifies the symbols used in the figure, and is sometimes placed in the figure itself. Although citations are generally used infrequently with figures and other
visuals, they are nonetheless important in providing complete information for
readers to accurately and conveniently assess the information being shown.

Finally, scientific journals may have in-house citation guidelines that differ
from those in standard manuals. Here are the detailed "Literature Cited" instructions from *Physiological and Biochemical Zoology,* with selected examples.

> Ex. 5.19
> Literature should be cited in the main body of the text by author name(s) and
> four-digit year of publication, with no comma separating the two. Multiple
> citations within a parenthesis should be made in chronological, not alphabet-
> ical, order, and separated by a semicolon. If two publications by the same
> author(s) appeared in the same year, the first should be designated by a low-
> ercase a, the second by b, and so on, following the date. Papers by one or two
> authors should be cited in the text by one or two names; papers by three or
> more authors should be cited by the first author's name followed by "et al.,"
> for example, Smith and Jones (1994a), but Johnson et al. (1995) for three or
> more authors. Bibliographic information should be given under Literature
> Cited, beginning on a new page and immediately following Acknowledg-
> ments. The listings should be double-spaced and arranged in alphabetical or-
> der. Publications by a single author should precede those by the same author
> with coauthors. Each reference should begin with the first author, name in-
> verted, with no comma separating last name and initials, followed by the
> other authors, with names not inverted. After the first line of each reference,
> succeeding lines should be indented. Manuscripts that have not been ac-
> cepted for publication must not be cited in the reference list, although the in-

formation can be mentioned in the text as unpublished observations or personal communications.

The name of a journal should be spelled out and not italicized; the volume number should also be set in standard type and not italicized. Italics should be used for scientific names. Full pagination should be given.

Examples:

Owerkowicz T., C. Farmer, J.W. Hicks and B. Branierd. 1999. Breathing under mechanical constraint: contribution of gular pumping to locomotor stamina in monitor lizards. Science 284:1661–1663.

Smith A.B. 1995*a*. The rise in blood glucose during hibernation of the golden headed plover *Dickus birdus*. Journal of Avian Metabolism 20:19–21.

Smith A.B. 1995*b*. The fall in blood glucose during hibernation of the golden headed plover *Dickus birdus*. Journal of Avian Metabolism 20:22–23.

Peck L. S. and L. Z. Conway. 2000. The myth of metabolic cold adaptation: oxygen consumption in stenothermal Antarctic bivalves. pp. 441–450 in E. Harper, J. D. Taylor, J. A. Crame, eds. Evolutional Biology of the Bivalve. Geological Society London.

Holyoak D. T. 2001. Nightjars and Their Allies. Oxford University Press, New York.[10]

The principles demonstrated in these particular citation examples, for a multi-authored journal article, two same-year articles by one author, a chapter in an edited book, and a single-author book, can be extrapolated to cover more complicated references if needed.

CITATIONS AS A REFLECTION OF PROFESSIONALISM

Wherever they are placed in a document, internal citations must be scrutinized to ensure that they match up precisely with the corresponding full citation in the notes or in a reference list. Beyond the importance of accuracy, Montgomery points out that citation in a scientific paper plays four different roles.

First, it offers accountability. It tells the reader that you are familiar with the most recent, significant literature in your area and that this literature has aided

you in your work. Second, citation is a way to outline a community of like investigators—a collegium, if you will. Third, citations are a tool by which you
express various degrees of agreement and disagreement toward the work of
others within this community: colleagues can be cited favorably ("the excel-
lent work of Barnes et al. 1987"), unfavorably ("Delpy [1994] failed to consider"), flatly ("has been the subject of numerous studies, e.g. Batts 1978;
Resin et al. 1983; Foresby 1985, 1992"), and in qualified fashion ("the work of
Jensen et al. [1998] requires further support"). Most documents employ several of these types—they are how scientist-authors rank their cohorts and competitors and position themselves toward them. Fourth, citation is also a way for
making certain claims to originality or, perhaps inadvertently, the very opposite.[11]

Montgomery makes these practices explicit to show realistically the true and
underlying complexity of how researchers may strategize bibliographically.
There are prominent cases as well of researchers who were perceived as omitting or downplaying credit in their publications, such as Darwin not crediting
the population theories of the mathematician Thomas Malthus or Watson insufficiently crediting Rosalind Franklin's crystallographic work. Documentation is a direct reflection of the professional standards and integrity of a paper's author. It is wise to select references carefully, not overloading a paper
with unnecessary citations simply because they are available. Conversely, citing that is too limited may have the effect of appearing reluctant to give proper
credit to the known work of others, as if one is trying to corner the credit. In the
end, attention to proper and balanced citation—despite the human tendency
toward bibliographic politics (including "buddy" citations)—is no less de-
serving of the honest and fair scrutiny given to one's own research, and ultimately researchers must live with both long into the future.

SCIENTIFIC VISUALS

THE IMPORTANCE OF SCIENTIFIC VISUALS

Scientific illustration has a long history, from ancient Greece to Renaissance Europe, with its astronomers, anatomists, and naturalists, to the Baconian experimentalists at the dawn of modern science, who by 1665 set forth a pioneer scientific journal, *The Philosophical Transactions of the Royal Society of London*—still published to this day, and still replete with visual depictions. The types of scientific visual representations and the manner of producing them have changed, of course, especially with the advent of electronic resources. It is commonplace for researchers to communicate scientific information visually. Even in the few instances when scientists write creatively for public readers, the concreteness of a visual image may serve to explain a point best. Some pages of Carl Djerassi's novel *The Bourbaki Gambit,* for instance, contain the DNA sketches his scientists use to illustrate the concept of a polymerase chain reaction (PCR); or, in the physician Michael Crichton's *Jurassic Park,* we are shown how his cloners keep close track of their dinosaurs with electronically tabulated and graphed data.[1] Although these particular visuals occur in an imaginary context, their function in communicating scientific information is standard.

Visual representations have highly formalized designs with standard parts, although (as with bibliographic styles) there is variability in prescriptions by editors, style manuals, and individual classrooms. While the emphasis here is on the use of visual elements in papers, the scientific illustrator Mary Helen Briscoe underscores their broader value in different forms of scientific communication: "A good illustration can help the scientist to be heard when speaking, to be read when writing. It can help in the sharing of information with other scientists. It can help to convince granting agencies to fund the research. It can help in the teaching of students. It can help to inform the public of the value of the work." Visuals are an effective medium for communicating scientific information because science is a highly visual activity and readers are readily engaged by graphical representation. Although technically visuals may not be considered a part of a paper's text, they are nonetheless an integral signifier of its intended meaning and must be treated with the same professional standards as any other uses of scientific language. Montgomery underscores the basic role of diagrams and charts: "The visual dimension to science forms a language all its own, a kind of pictorial rhetoric, if you will. By this I mean that graphics are often much more than a handmaiden to writing. They don't just restate the data or reduce the need for prose, but offer a kind of separate 'text' for reading and interpretation. . . . You will find that they tell their own story, in some manner parallel to that of the writing, but in other ways different, enriching, though also with notable gaps." The validity of this assertion can easily be verified, Montgomery suggests, by isolating the figures of any amply illustrated paper, lining them up in their original order, and noting how they tell their own story in their peculiar relation to the textual narrative. Before examining a few examples of different kinds of illustrations used in scientific papers, it will be appropriate to start with some basic questions: What types of visuals are used in scientific papers and for what purposes? How are scientific visuals planned and designed to be readable as well as understandable by the paper's intended audience?[2]

PURPOSES SERVED BY VISUALS IN SCIENTIFIC PAPERS

Illustrating scientific papers means using tables and figures to communicate information when words alone would not do so as clearly, fully, or convincingly. As Wilkinson points out succinctly, "In a scientific paper, any visual representation that is not a table is called a figure, which may consist of a single illustration or several."[3] Tables display numerical or verbal information in columns and rows or in lists, while figures—such as flow diagrams or bar graphs—are pictorial and more dynamic. Visuals typically are placed in sections of a scientific paper that present the findings, whether from bibliographic research (as in most college reports) or from experimental research that produces new information worthy of publication in a journal. In an experimental article, visual elements are most often placed in the methods and results sections. The purposes served by visuals (or synonymously "graphics") are diverse. They are used to:

- display experimental data;
- provide material evidence of newly discovered entities;
- show objects, features of structure or function, or natural phenomena;
- demonstrate physical, temporal, or spatial interrelationships;
- illustrate processes, concepts, or new theoretical models.

Innovative equipment or techniques, or results involving changes in physical appearance, may require photographic evidence. Visually observable experimental outcomes, or measurements that generate data in graphical forms, as in chromatographs or computer-generated formats, also may call for their being shown to readers. Or when methodologies generate large amounts of numerical information, visuals permit a condensed presentation that also allows readers to focus more readily on specific features, relationships, or trends in the results. In short, graphics provide scientific information more clearly, concretely, concisely, and convincingly than would otherwise be possible.

PLANNING AND DESIGNING VISUALS

Long before the first draft of a paper is attempted, thoughts regarding the selection and preparation of visual representations will have occurred during the research process itself. This again is due both to the visual nature of scientific work and to the forms in which data are generated. Among the basic considerations that must be addressed as each graphic is being conceived, planned, and developed are:

- When is a visual really needed?
- Which type is most suitable for the particular information?
- Is the visual adapted to its intended viewer?
- Is the image designed effectively for its medium—for example, a paper, a slide, a poster?

- What is the relationship between figure and text?
- Does the visual follow expected conventions of format and labeling?
- Is the graphic incorporated properly within the paper's text?
- Is the visual's format or amount of information accessible to the reader?

Decisions regarding the choice, preparation, and textual integration of visuals can begin during data collection and note taking, or later at an outline stage for a paper. Whether early or once the experimental work is completed, researchers can draw preliminary sketches that include notations regarding such features as scale, headings, and labels. Once a decision is made to use a particular illustration, it is necessary both to assess its clarity of depiction and to place it in the appropriate part of a paper. The introduction to a paper seldom includes visuals. Illustrations that show a technique or procedure belong with the description of materials and methods, those that present data naturally go in the results section, and those that synthesize ideas and represent concepts or models typically are placed in the discussion and conclusions. All visuals should be numbered consecutively (tables and figures separately) and the paper's text should refer to each one. An important part of visuals is their title or legend, which not only explains what is being shown but also connects it to the paper's text. In practice, titles, legends, and captions vary widely, with some representations having a short sentence-style title and others a legend running from several sentences to the extreme end of hundreds of words. A few selected examples of tables and figures will suffice to illustrate their range of design and purpose. For continuity and focus, the examples offered here are taken primarily from the alcohol studies area.

PREPARING TABLES

"Tables, like dictionaries," to borrow Briscoe's apt analogy, "are indispensable in our lives."[4] We use tables of contents, of measurements, tides, sports data, pedigrees, genealogies, genetic crosses, financial amortization, mathematical randomization, chemical elements, and public transportation schedules—even calendars have tabular form. In scientific papers, tables summarize and group information to show raw data, calculations, or experimental results in an easy-to-follow manner that facilitates comparison and verification by readers. Your readers will be grateful when your paper's tables are designed to provide the path of least resistance—even issuing an appeal—to comprehending and using its array of scientific information. What are the elements of an effective table?

First, whether a table displays numerical data from experimental results or some kind of verbal listing, it must do so not only with unequivocal accuracy but also for good scientific reason. Presenting some types of information or experimental data in a paper's text may be either too cumbersome, harder to follow, or less amenable to demonstrating significant relationships or patterns in the research findings.

Second, a table's logic and simplicity of design, including how its labels are worded and placed, should make it easy to follow and use. The information of key significance should be readily apparent, so that readers are guided through it vertically or horizontally, with minimal risk of misreading, ambiguity, or even misinterpretation.

Third, a table should be complete enough to be self-contained, but also linked to the text in an apparent way. While readers should be able to comprehend the table's information with minimal reference to the paper's text, the author should make the connection between the two clear by referring to the table at an appropriate time (for example, when the subject of the information it contains first arises).

Fourth, tables should be designed with a conventional format that identifies and circumscribes their purpose. Also important in the design of tables is spatial economy, appropriate textual placement, and consistency in their features when used multiple times throughout a document. As we begin to look at examples, it will be helpful to note that tables are designed with the following kinds of typical parts:

- Table number designation
- Title
- Columns, with headings
- Rows, with row headings in the "stub" (left-most) column
- Field (cells containing data, or listed items, collectively)
- Lines, or rules (for columns, rows, subheadings, spanner heads, textual separation)
- Notes or references (footnote, headnote, sourceline)

The key content of the table is the information that is placed in its field, whether of a numerical or a verbal nature. However, in addition to making evident the compelling significance of the information in its field, a table must

Table 6.1   ICD codes for alcohol-related causes of death

Cause of death	ICD-6–7	ICD-8	ICD-9
Liver cirrhosis*	581	571	571
Alcoholic diseases of the liver	—	—	571.0–571.3
Alcoholism/Alcohol dependence syndrome	307	303	303
Alcoholic psychosis	322	291	291
Alcohol poisoning	E880	E860	E860
Alcohol abuse	—	—	305.0
Alcoholic cardiomyopathy	—	—	425.5
Alcoholic gastritis	—	—	535.3
Alcoholic polyneuropathy	—	—	357.5

*Chronic liver diseases since ICD-9
Dashes indicate that revisions of the ICD (International Classification of Diseases) prior to the ninth (ICD 9) did not include a code for these alcohol-related causes of death.
Source: European Journal of Population 18, no. 4, 2002, 310, © Springer

also be planned so its users can clearly see the interrelationships among the various features, categories, and patterns of the scientific information that it displays.

The first two tables shown here are from a study on alcohol drinking pat terns in human populations in Europe. The tables are from different parts of the same paper and differ in their design's sophistication. The first example, from the paper's section on data and methods, is a list (Table 6.1).[5] Its design is uncomplicated, with few rules, columns, and rows, as well as simple headings. Its field contains a modest amount of information: alcohol-related causes of death and their codes in four revisions (6th–9th) of the International Classification of Diseases (ICD). The paper's text provides effective linkage to the table by describing the study's population (mostly European Union countries) and the period for which data was collected (1950–1995).

The next table, from the same study's section on results, is designed more elaborately, with layered and split headings and a more sophisticated field (Table 6.2). The table's title is accompanied by a key for the number-coded "AAA-mortality" categories and a note on the study period. The field itself is organized by two-tiered headings for rows (region, countries) and split columnar headings (ICD mortality by gender). The paper's text assists readers further by pointing to the location of regional averages in the table's field (first data lines by row headings). The two tables shown together with several oth-

Table 6.2 Alcohol-related deaths in Europe for select ICD categories, expressed as a percentage of total alcohol-related deaths. Categories shown are Alcoholism (303), Alcohol psychosis (291), Alcohol poisoning (E860), and Other, which includes alcohol abuse (305.0), alcoholic cardiomyopathy (425.5), alcoholic gastritis (535.3), and alcoholic polyneuropathy (357.5). Percentage figures are annual averages for 1987–1995.

Country	Men				Women			
	303	291	E860	Other	303	291	E860	Other
Northern Europe	50	4	36	10	44	3	46	7
Finland	13	6	61	20	12	4	70	14
Norway	72	3	21	4	63	2	35	5
Sweden	64	3	27	6	58	2	35	5
Central Europe and the British Isles	66	5	9	20	66	3	13	20
Austria	85	2	0	13	87	1	0	12
Belgium	80	8	2	10	78	6	2	14
Denmark	84	1	14	1	74	1	24	1
Ireland	68	4	28	0	62	1	37	10
Netherlands	51	9	3	37	54	11	2	33
UK	37	3	17	43	40	1	20	39
West Germany	60	5	1	34	66	3	1	30
Southern Europe	78	13	3	7	76	12	3	9
France	82	9	0	9	87	6	1	6
Greece	87	8	1	4	63	20	0	17
Italy	83	8	2	7	81	8	3	8
Portugal	66	24	9	1	82	15	3	0
Spain	72	15	1	12	68	12	7	13
All countries	67	7	12	13	65	6	16	14

Source: European Journal of Population 18, no. 4, 2002, 314, © Springer

ers in the paper clearly tell their own supertextual story, from the studied diseases to the list of countries and observation periods to the study's results presented by gender and ICD mortality category.

The following example, from another paper's results section, uses tiered and split headings in presenting alcohol consumption data, but with additional features (Table 6.3).[6] Its title is short and simple, with contextual information placed at the bottom (versus the top in Table 6.2) in a footnote, plus a key for

Table 6.3 Binge drinking among adults in the United States who consumed alcohol, 2001*

Characteristic	Males (n = 57,654)		Females (n = 46,811)		Total (n = 104,465)	
	%[†]	Rate[‡]	%[†]	Rate[‡]	%	Rate
All respondents	35.9	20.1	15.7	5.8	26.8	13.7
Age						
18–20	61.1	39.0	37.7	17.6	51.3	30.0
21–25	61.9	38.7	32.0	12.5	48.6	27.1
26–34	44.2	20.8	20.2	6.5	34.1	14.8
35–54	33.3	17.9	13.5	4.7	24.2	11.9
≥ 55	15.0	10.4	4.7	1.8	10.2	6.4
Race/ethnicity						
White	34.6	19.3	15.6	5.6	25.8	13.0
Black	33.4	20.7	14.1	5.1	24.4	13.4
Hispanic	45.3	23.5	18.1	7.0	35.1	17.3
Other	37.2	24.1	15.3	6.6	29.0	17.6
Education						
Some high school	45.3	29.8	23.1	10.1	37.5	23.0
High school graduate	41.9	26.3	16.6	6.2	30.7	17.4
Some college	38.5	21.6	17.8	7.0	28.3	14.4
College graduate	27.0	11.9	11.7	3.6	20.3	8.2
Alcohol intake[§]						
Moderate	30.1	9.5	11.3	2.5	21.6	6.4
Heavy	88.2	113.8	59.6	37.9	76.0	81.9

*Binge drinking is defined as consuming ≥ 5 alcohol-containing drinks on 1 occasion.

[†]Percentage of American adult drinkers who had at least 1 binge-drinking episode in the past 30 days.

[‡]Number of episodes of binge drinking per person per year (among drinkers for given demographic group).

[§]Moderate alcohol drinking is defined as consuming an average of ≤ 1 alcohol-containing drink per day for a woman or ≤ 2 for a man, and heavy alcohol intake as consuming an average of > 1 alcohol-containing drink per day for a woman or > 2 for a man.

Source: JAMA 289, no. 1, 2003, 73, © American Medical Association

Table 6.4 Relation of behavioral depression, as indicated by behavior in the "forced swim" and "stress–open field" tests, to voluntary alcohol consumption in genetically defined rodent strains

Strain	Forced swim test	Stress–open field test	Voluntary alcohol consumption
Flinders sensitive rat	?	+	0
P rat	0	—	+ +
Fawn-hooded rat	?	+	+
C57 mouse	+ +	+	+

Key:

? response unknown

+ sensitive to behavioral depression; voluntary alcohol consumption

+ + very sensitive to behavioral depression; high levels of voluntary alcohol consumption

0 no sensitivity to behavioral depression; no voluntary alcohol consumption

— not tested

Source: Alcohol Research and Health 26, no. 3, 2002, 235

the symbols used next to the paired column subheads for each sex († and ‡) and the "Alcohol intake" stub entry (§). The column headings also provide parenthetical data (number of drinkers). In addition, the table's readability is enhanced by the following: spanner rules beneath the column headings (Males, Females, Total) to clarify the relation of their subheads (% and Rate) to them; extra rules to separate groups of rows; and indentations for the row subheads (e.g., each age group).

The final example, from a research update in a journal, summarizes reviewed studies in a table using symbols that are explained in a key at the bottom (Table 6.4).[7] Because of the symbols in each cell, the row and column headings stand out.

As simple as the concept of a table may be, discussion about its design features can become rather technical on any number of aspects. Consider the following prescriptive language in the CBE manual on a table's alignment, for instance: "When a row heading in a single-spaced table carries over to a 2nd line, that line should be indented; entries needing only a single line opposite a multiple-line row heading should be placed opposite the 1st (unindented) line of the heading. Occasionally, circumstances (for example, text tables with entries of several lines each) or aesthetics will dictate that a table be set with blank lines between rows. In such cases, carryover lines in the stub need not be indented, and single-line entries opposite a multiple-line row heading should again be set opposite the 1st line of the heading."[8] Notwithstanding the descriptive usefulness of such language for setting and codifying standards, ultimately a table's visual effectiveness depends on its clearly interconnected features working smoothly together. These elements include: a helpful title and headings; suitable and sufficient information in the field; spatial economy and clarity in design and layout (field, spacing, headings); headnotes or footnotes for glosses or keys; rules that help the reader comprehend information at a glance; and evident linkage to the main text.

Another consideration in a table's use, as with any other type of visual aid, is the manner of its textual incorporation: Will it be placed in a single column of text or across columns? Text-wrapped? Vertically? Broadside? Split-page? Scientific style manuals are in general agreement regarding the elements of effective tabular design, with minor variations in such features as titles (e.g., sentence versus headline style, centering), or incorporation across columns in a paper's text. Authors of articles also must adhere to individual publishers' guidelines for designing and submitting tables. For college papers, students may have to work within the tabular options available in networked software (such as Microsoft or Corel products). Whatever the design features, a paper's readers must feel that a table's information merits being fussed over, that the table itself is not expendable. For publishers, tables (especially of elaborate design) must also be worth their added reproduction costs, and therefore they should be kept as simple as possible.

PREPARING FIGURES

Visual representations other than tables are typically called (and identified as) figures. This dichotomy leaves the possibilities wide open—virtually any data or image of scientific significance can be worked into a figure with standard design features. Scientific figures may be of the following types:

- Photographs (of objects, organisms, microscopic images, medical conditions, for example)
- Bar graphs (showing scalar data, comparisons over time across different variables)
- Line graphs (mapping events over time, frequency or distribution curves)

- Point or dot graphs (representing nonscalar data or variables, scattered data)
- Circular or pie graphs, also known as pie charts (comparing sliced data within a whole)
- Diagrams (of equipment, flow schemes, molecules, conceptual models)
- Charts (showing genetic or organizational relationships, or maps)
- Drawings (freehand, line, mechanical)
- Combination or multitype figure (such as a photo with a graph)

Students preparing figures for college papers typically rely on the capabilities of the software available on their campus network for creating such visuals as graphs, charts, and drawings. This is especially so given the often prohibitive cost of purchasing specialized graphics software, such as OR-TEP for molecular crystal structures, CAD-CAM for engineering designs, and Sigma Plot or Harvard Graphics for more general visual applications. On the other hand, the declining cost and wider accessibility of digital technology has facilitated the incorporation into papers of photographic and scanned images. Beyond an awareness of the range of resources and options available for designing and incorporating different types of figures, the familiar kinds of questions arise that apply to all visuals: When should they be used? What standard parts are needed? How can they be most effectively designed and placed? Could the same information be represented in different forms? Pie chart? Bar graph? Table? Sets of data in the columns of a table may be better shown as a series of pie charts, for instance, which will allow emphasis of certain proportions with slices pulled away or with color schemes.

As with tables, presenting information in a figure must be done for compelling and evident scientific reasons—that words alone will not fully serve—and in the interest of making the main argument and experimental results clearer, more complete, or more convincing. Once a studied decision is made to use a particular figure, care also must be taken to include the following kinds of standard parts in its design:

- Figure and number designation
- Title
- Labels for internal parts (e.g., lines, tags, arrows, letters, names)
- Caption or legend (explanatory phrases or sentences)
- Key (for symbols)
- Notes or references (footnote, sourceline)

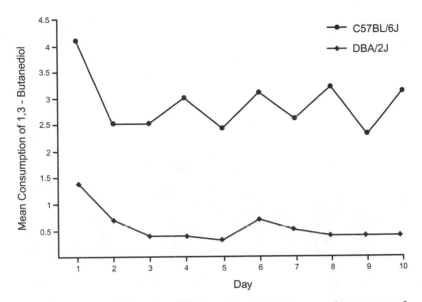

Figure 6.1 Line graph showing differences in alcohol consumption patterns of
high-drinking (C57) and low-drinking (DBA) mouse strains

Considering the energy and time that it can take both to prepare and to view
them, figures must be worth their keep in both their scientific value and their
appearance. Visuals perceived to be frivolous or unhelpful may call into question the paper's overall authority and reliability. Given the great diversity of
information, ideas, or images that may be incorporated into scientific papers
as figures, the examples that follow are intended only to provide a sense and
sampling of their range in purpose, content, and design.

The first figure is a line graph with just two sets of data from an alcohol
study (described in Chapter 2), intended to show that two mouse strains—C57
and DBA—have non-overlapping drinking levels of 1,3-butanediol over a 10-
day test period (Figure 6.1). Labels are placed near each line, or elsewhere in-
side the graph, although this can be cumbersome if there are too many labeled
lines that clutter the figure. An alternative is to place a key below the figure to
identify each line. When there is much data to convey, using a series of graphs
with fewer lines is better than one crowded figure. When the same data can be
represented as either a table or a line (or bar) graph, the advantage of the graph
is that patterns can be seen more readily, even without immediately knowing
the exact values. The symbols used in line graphs (circular, triangular, rectan-

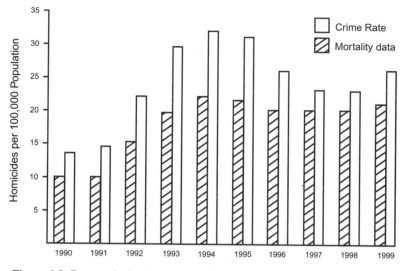

Figure 6.2 Bar graph showing Russian homicides per 100,000 population
according to crime and mortality data, 1990–1999 (*American Journal of
Public Health* 92, no. 12, 2002, 1924, © Springer)

gular, or diamond shapes) can be used to allow readers to see related plots or to
tell apart contrasting types of data.

The bar graph shown here, from the results section of a Russian study on the
relation of alcohol consumption to violence, illustrates how multiple sets of
data can be compared side by side (Figure 6.2).[9] The text of the paper explains
that the pair of bars for each year throughout the 1990s shows the substantial
discrepancy between homicide data officially *reported* (Ministry of the Inte-
rior) and homicide data *recorded* (Ministry of Health). Different fill patterns
can be used to distinguish bars for each type of data, as done here by white versus a diagonally-lined pattern, and some bar graphs use color (a more expensive option to reproduce). Using color for college papers is not a major expense issue, but papers intended for publication should use simple black and
white patterns: diagonal lines, dots. Shading should be restricted to one shade
of medium gray (and no shading of the background) for the best reproduction
quality. Bar graphs also may be oriented horizontally (useful for long bar labels), or they may represent data in more sophisticated ways, such as in varie-
gated stacked patterns separated within single bars.

Another type of figure is a scatter graph, such as the example here from a
study of two lizard populations showing the relation of the animals' body

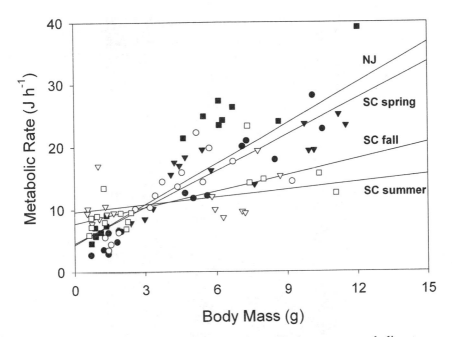

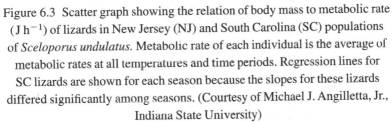

Figure 6.3 Scatter graph showing the relation of body mass to metabolic rate
($J\,h^{-1}$) of lizards in New Jersey (NJ) and South Carolina (SC) populations
of *Sceloporus undulatus*. Metabolic rate of each individual is the average of
metabolic rates at all temperatures and time periods. Regression lines for
SC lizards are shown for each season because the slopes for these lizards
differed significantly among seasons. (Courtesy of Michael J. Angilletta, Jr.,
Indiana State University)

mass, on the x-axis, to their metabolic rate on the y-axis (Figure 6.3).[10] In this
case, instead of a key, the lines are individually labeled to distinguish the seasonal data for each population. The caption assists readers by explaining how
metabolic rate was derived and why the graph shows regression lines for each
season for the lizards from South Carolina. In scatter graphs, sometimes loga-
rithms of the data are plotted instead of the raw numbers, to avoid skewing effects from a few extreme values and to allow for a more normalized distribution. One issue with logarithmic plots, notes Katz, is that they "downplay
differences between large values; we may not be able to perceive trends hid-
den in the high end of logarithmic graphs, or we may overemphasize variations exposed at the low end of logarithmic graphs."[11] For both line and scatter graphs, a scale must be used that accurately represents the data's magnitude

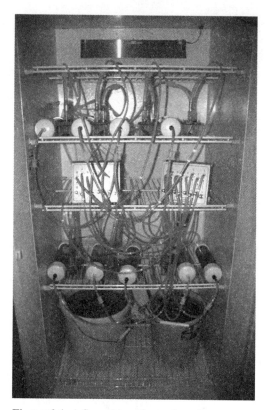

Figure 6.4 A flow-through respirometry system
used to measure the metabolic rate of lizards
(Courtesy of Michael J. Angilletta, Jr.,
Indiana State University)

and trend. Because graphs may have inherent limitations in displaying some types of data unambiguously—in their relationships or patterns, for example —it may be necessary to accompany some graphical representations and conclusions made from them with numerical or statistical analyses.

Sometimes a photograph is necessary, for example when the author wants to show the apparatus used in an experiment (Figure 6.4). The example here shows a flow-through respirometry system that was used to measure the metabolic rate of lizards in the study described above, for a seminar by a doctoral student at the University of Pennsylvania. Photographs are commonly used in scientific presentations and journal articles, and they sometimes are accompanied, especially in a complex illustration, by labeling of their parts.

Occasionally one may need to combine several different types of illustrations in one figure—for example, microscopic images of chromosomes, a diagram of the double-helical structure of DNA, and letter sequences used to represent genetic markers, used in this case to suggest how genetic analysis could be applied to the study of alcoholism (Figure 6.5).[12] Multipart figures, with individual letter designations (each part labeled A, B, and C, for instance), are used in scientific papers with some regularity, often with a series of a single figure type such as bar or line graphs. The caption to Figure 6.5 contains in effect three titles (italicized here for emphasis), one following each letter designation for the three parts of the figure. While the captions in Figures 6.3 and 6.5 may seem cumbersome in their length and degree of detail, captions in published papers may be considerably longer, even in the hundreds of words, and take up much more space than the figure itself.[13] Long captions are appreciated by readers who prefer to focus on the representation of results or concepts in a paper's visuals, rather than necessarily wading through the entire paper to find that information.

Figures are also useful for showing complex relationships or processes in schematic form to make their essential points easier to grasp. One example might depict the possible intergenerational effects of alcohol dependence on psychiatric disorders like depression, using ovals, boxes, and arrows to show a hypothetical genetic process (Figure 6.6).[14] The figure's lengthy title also serves as an explanatory caption, along with a note that glosses the use of question marks to indicate uncertainty regarding a relation between components of the model. Schematics are used frequently in scientific writing, not only to suggest theoretical models that explain observed phenomena but for a range of other purposes, such as showing electrical circuitry, the stages of a process (such as water treatment, cellular division), and biochemical mechanisms.

A similar example is a procedural flow chart, which is useful for showing the steps followed in an experiment. Researchers often use complex processes, such as combining two techniques of biochemical analysis—gel electrophoresis and mass spectroscopy—to identify a sample protein fraction (Figure 6.7).[15] The caption for this example generally is effective and restrained in its detail, with only one explanatory sentence following the title line, despite its unwieldy noun cluster—"matrix-assisted laser desorption/ionization time-of-flight mass spectroscopy"—the name of a technique that may not be readily amenable to rewording. Flow charts may also be circular,

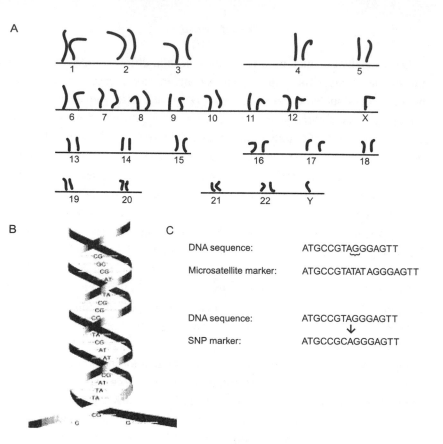

A

1	2	3	
		4	5

| 6 | 7 | 8 | 9 | 10 | 11 | 12 | X |

| 13 | 14 | 15 | 16 | 17 | 18 |

| 19 | 20 | 21 | 22 | Y |

B

C

DNA sequence:	ATGCCGTAGGGAGTT
Microsatellite marker:	ATGCCGTATATAGGGAGTT

DNA sequence:	ATGCCGTAGGGAGTT
SNP marker:	ATGCCGCAGGGAGTT

Figure 6.5

A: *A set of human chromosomes as seen under a microscope, containing 22 chromosome pairs (ordered according to size) and 2 sex chromosomes.* In this case, the chromosomes were obtained from a male, as indicated by the presence of an X and a Y chromosome.

B: *The structure of DNA.* The DNA molecule is composed of two strands of building blocks that interact with each other. Each building block contains a chemical called a base. There are four bases, adenine (A), cytosine (C), guanine (G), and thymine (T), in a sequence of paired bases. The base A on one strand always pairs with T on the opposite strand, and G always pairs with C. The sequence of these bases encodes the genetic information.

C: *Microsatellite and single-nucleotide polymorphism (SNP) markers.* Microsatellite markers are short sequences of two to four bases (in this example, the bases T and A) that are repeated several times. The number of repeats differs among individuals, creating many different versions (i.e., alleles) of the marker for genetic analyses. For SNP markers, only a single base differs between individuals (in this case, the base T is changed to a C); thus, there are only two possible alleles of the SNP.

(*Alcohol Research and Health* 26, no. 3, 2002, 173)

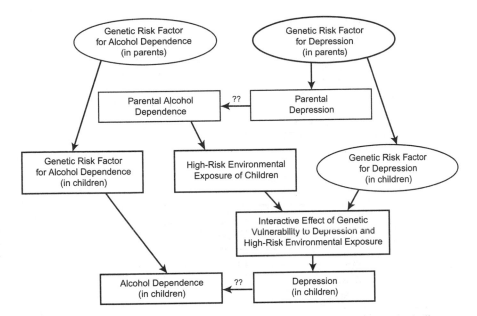

Figure 6.6 Schematic model showing how intergenerational processes, including genotype × environment interaction effects, may contribute to the development of alcohol dependence and comorbid psychiatric disorders, as illustrated by the example of depression. The question marks indicate uncertainty about whether depression directly affects the risk of dependence. (*Alcohol Research and Health* 26, no. 3, 2002, 197)

showing continuous feedback loops such as those depicting mechanical or biological systems.

WEIGHING OPTIONS AND MEETING VISUAL EXPECTATIONS

The options for communicating scientific information visually are diverse, and careful consideration is required in the selection and design process. Choices must be carefully weighed with regard to their ultimate practical value to readers in their comprehension, interpretation, and evaluation of the data or concepts that they are being asked to view, decode, and apply in their own work. Visuals are not merely an afterthought or an expendable aesthetic in a scientific paper. Once options have been considered and decisions have been made regarding their use, form, and placement in a paper, visual elements must be treated as seriously as any other form of scientific language for any clear, reliable, and convincing scientific exposition. Given the diversity of

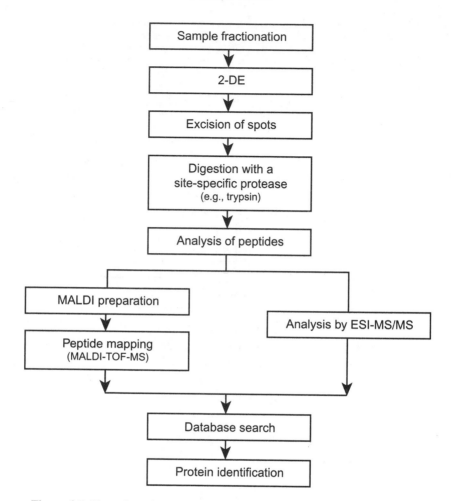

Figure 6.7 Flow chart showing the process of protein identification through a combination of two-dimensional gel electrophoresis (2-DE) with mass spectroscopy (MS). If the matrix-assisted laser desorption/ionization time-of-flight mass spectroscopy (MALDI-TOF-MS) approach does not result in protein identification, additional analyses, such as electrospray ionization (ESI) combined with at least two steps of MS, may be used. (*Alcohol Research and Health* 26, no. 3, 2002, 222)

scientific specialties, with all their possible forms of graphical representations, the intention here has been far from cataloguing and illustrating every type that researchers may incorporate into papers.[16] Moreover, beyond the basic or standard appearance of tables and figures, there are always creative variations in design by individual authors, such as in labeling, fonts, color, combinations of visual types, and conceptual or theoretical representations in models. It is also likely that the development of new approaches or technologies for scientific observation and measurement, perhaps applied to yet unknown phenomena, will bring accompanying forms of innovative graphical representation or display. With the professional maturation that long-term research and authorial experience confer, scientist-writers also gain opportunities to refine their judgment and decision making as to when and how they present information in their papers visually. In any case, aside from the available or original options for communicating data or ideas graphically, the criteria of accuracy, clarity, completeness, consistency, necessity, and readability will remain as overall standards of professionalism in scientific visuals.

7

SCIENTIFIC PRESENTATIONS

I myself have now for a long time ceased to look for anything more beautiful in this world, or more interesting, than the truth; or at least than the effort one is able to make towards the truth. I shall state nothing, therefore, that I have not verified myself, or that is not so fully accepted in the textbooks as to render further verification superfluous. My facts shall be as accurate as though they appeared in a practical manual or scientific monograph, but I shall relate them in a somewhat livelier fashion than such works would allow, shall group them more harmoniously together, and blend them with freer and more mature reflections.
—Maurice Maeterlinck, *The Life of the Bee*

THE PROFESSIONAL VALUE OF SCIENTIFIC PRESENTATIONS

The apiarist Maurice Maeterlinck's promise to be both truthful and lively in his 1901 personal essay on bees reads almost like a ceremonial oath that all scientific presenters could just as well recite today at the podium of a conference room. Indeed, any form of scientific presentation—whatever its setting, audience, or formality—carries with it the same professional responsibilities that apply to *writing* science, but adds a dimension that is very different: It is a live performance, before listeners who are not always obligated to remain in their seats or may tune you out even if they are. Moreover, since the audience will receive the information only once, without having the luxury of a print copy to read and reread at will, the presenter is compelled both to get it right the first time and to say it in an engaging way. Oral scientific presentations can take various other forms besides that of conference papers delivered to professional peers. Scientists may provide oral testimony, for instance, accompanied by written statements, in criminal cases (DNA, biomedical effects) or in congressional hearings (cloning, energy policy). Given the employment of today's scientists outside university settings, in the worlds of business and government, opportunities for oral presentations have only grown and diversified. A researcher for a government agency like the Centers for Disease Control and Prevention may find herself presenting a statement in a press conference, for example, or one working for a biotechnology corporation may need to make marketing presentations (pharmaceuticals, laboratory techniques). Scientists employed in universities give lectures, seminars, and even talks at local libraries or for special-interest groups to serve their community. No matter its form, even in informal exchanges with labmates, sharing information orally is a valued and regular aspect of modern scientific work. In our undergraduate and graduate science curricula, students eventually are called upon to give a talk to their peers or to a more diverse collegiate audience. The frequency of giving talks tends to rise with the upward spiral of scientific training and professional maturation.

Given their regularity and special role as a professional forum, conference-style presentations (practiced in precursor forms by students) are the focus here, with attention as well to poster presentations. Although talking is the basic component in presentations, both oral and poster presentations of scientific information have substantial written and visual components that contribute critically to their effectiveness. We may begin by asking: What do scientific presentations offer beyond what the printed word does?

UNIQUE BENEFITS OF ORAL PRESENTATIONS

While a print medium, such as a journal, is a standard way to make a researcher's work public in its most extended and detailed form, oral presentations at professional gatherings constitute a primary and dynamic process of "live" scientific exchange that both presenters and listeners find uniquely beneficial in a number of ways.

First, receiving the information directly—in person, from the actual researcher—is a powerful affirmation of the authority and competence of the scientist who did the work and who now stands ready to defend its significance. Audience members can ask questions, seek clarification, and afterward meet with presenters for further exchanges on specific points of mutual interest. In this informal way, researchers can engage in the personal dialectic that is so essential to scientific inquiry and share information that is not so accessible in print versions of their work. Behavioral observations with animals can be described more closely, failed or modified experimental protocols can be shared, and concepts or theories can be cross-critiqued. Such exchanges may not only provide practical scientific insights but also lead to new professional relationships or opportunities for collaborative work.

Second, there is an immediacy to both disseminating and receiving freshly completed work, without the delay of its prospective print form. Researchers get to announce their findings right away and attendees get to apply the new insights that much sooner. A print version of the work, typically in a field-specific journal, could take many months to appear. Therefore, an oral presentation is a way to offer a preliminary, though nonetheless self-contained, announcement of findings.

Third, presentations can make use of ways to communicate information that are more dynamic than a hard-copy print. These include presentational software (such as Microsoft's PowerPoint or WordPerfect's Corel Presentations), electronic projection screens, film, and demonstrations with models or other visual aids. The human presence and voice itself add a dimension that tends to be underappreciated by speakers—vocal inflections, facial expressions, and body language showcase one's own sense of engagement with the work being presented. At the same time, presenters can read the reactions and background of audience members and if necessary make spontaneous adjustments for their clarity and understanding.

Fourth, presentations provide special opportunities for the public media to find newsworthy scientific stories. That this audience may not figure prominently in the presenter's mind does not diminish its importance as a resource for translating and disseminating the presented work to the wider public. It is one way, therefore, for researchers to live up to a professional commandment, to use Bentley Glass's term, that dates back to Francis Bacon's ideal of science for the ultimate good of the citizenry.[1] Today this professional role is magnified, given the billions of tax dollars granted annually to researchers together with the great power of scientific advances to transform public life (through bioengineering, nuclear energy, or nanotechnology, to name just a few). Public accountability means, therefore, accessibility of scientific knowledge to taxpayers who must vote on public policy issues requiring scientific insight and foresight.

Although these are all good reasons for researchers to value giving conference talks, presenting one successfully means much more than merely going through the motions and mechanically reciting a paper's parts (in IMRAD style, for instance). Just as with a research report or a journal article, there are certain key considerations that go into making an effective presentation. Giving a scientific talk requires close attention to certain basic elements—temporal, vocal and auditory, written, and visual—that must work harmoniously together.

TIMING

The temporal element in a presentation has three senses: time allotment, time distribution, and pacing. Each of these factors makes a difference in its own way. The time allotted for a conference-style presentation—whether 8 minutes, 12 minutes, or 15 minutes—typically is a firm given that is tightly controlled by a session coordinator (or, in a classroom, by an instructor) and will determine the length of a presentation's text. This naturally means that the number of words spoken must fit the given time frame. In a 10-minute talk, for instance, about six pages of text can be delivered effectively. It is always awkward and uncomfortable when a speaker's pace suddenly quickens as time is running out, taxing the patience of listeners. It is even worse professional manners to exceed the time limit and thereby frustrate the expectations and plans of others—not only the speakers who follow but the attendees who have scheduled their own day.

Managing time in a presentation also means determining its apportionment among the paper's parts, from introducing the subject and its purpose to presenting and discussing the findings, and then concluding. Most of the allotted time will be given to reporting and discussing the findings, and material must be parceled out temporally, with relatively more time needed for especially complex information, points, or arguments and their accompanying visuals. As in written papers, the time spent introducing and concluding a presentation, versus on presenting the findings and explaining their significance, is proportionally shorter. Background and citations are pared to a minimum.

Time also must be left and efficiently utilized—two or three minutes between presentations—for a question-and-answer period.

Effective pacing takes into account how quickly or slowly particular kinds of information should be delivered. Presenting experimental results, for instance, requires a slower pace of speaking compared with, say, introducing the research, referring to background work, or describing methodology. The pace must be appropriately slowed and measured when listeners also have to view, read, and process projected information—images, graphs, numbers, labels—in order to keep up with the scientific narrative's building significance. A presenter may read the faces of listeners to gauge whether the pace could use adjustment at any given moment in the narrative.

How a presenter handles the various time considerations is an important determinant of a talk's degree of success. Timing and pacing can contribute to a rapport between speaker and audience. Fundamentally, good time management is both a reflection of your preparation and a courtesy to your listeners.

SPEAKING

Like a speaker's time management, vocal qualities can work either to invite attendees to participate in the unfolding story and to keep listening, or to alienate them. How vocal qualities are used will affect how listeners process and respond to the communicated information. Parallel with the notions of readability and viewability for written and visual communication, respectively, we may also ask whether a talk is optimally "hearable." Are the words being enunciated clearly and distinctly? Is the voice a flat monotone or does it exhibit a personal and enthusiastic engagement with the work being presented? Are the speaker's pitch and tone inviting or distancing? Does the speaker use vocal inflections to signal or to emphasize key or transitional points in the scientific narrative? Vocal qualities, along with nonverbal cues and body language, can have a substantial effect on either keeping or losing an audience.

"Of all the monsters of science fiction," writes Medawar in his little volume *Advice to a Young Scientist,* "the Boron is that which arouses the greatest dread—anyhow at scientific conferences."[2] Even a well-organized paper, if presented in an excruciating monotone, can send its listeners toward the exit seeking more captivating storytellers. Those who remain, whether by charitable politeness or special interest in the subject, may nonetheless think twice about attending that presenter's talks in the future. Worse yet, much of the talk will have been lost on hostile ears and napping minds. For listeners to hear the story out, the speaker must be prepared to help them through it by bringing the work to life before them.

WRITING

"Under no circumstances whatsoever should a paper be read from a script," Medawar exhorts.[3] This sound advice does not negate the likelihood, however, that some form of writing—or, for the audience, reading—will contribute substantially to the effectiveness of a scientific talk. Writing has a place in both preparing and delivering an oral presentation. Very early on, even as a researcher gathers information, the rough parameters of a presentation, including visual options, may begin to be envisioned in the research notes themselves. Once the research is completed, direct planning may begin by outlining. For experimental papers, especially in biology and chemistry, an outline may follow the IMRAD model. For theoretical or review papers, the outline's structure will depend on the speaker's particular purpose or argument. An outline for a presentation, like one for a written paper, should include the key points in the major parts of the paper, from background and thesis to methodology, findings, and conclusions. Other details may include references to the work of others and the sequencing of visuals. Once these key points are in place in the outline, a full script may be written out—for rehearsal purposes only—and later distilled into notes for the actual presentation.

Another important written item is the presentation abstract, which has two key and distinct functions. First, it is submitted by a prospective presenter as a response to a call for papers. All of the abstracts then go through a competitive selection process to determine which of the submitted papers will be included in the conference program. Once a paper is accepted for presentation, the abstract plays a second role: it is included in a pre-conference publication that is given to all participants as a guide for deciding which sessions and whose talks to attend. It is not difficult to imagine, therefore, how much of a difference an effective abstract can make.

After preparing the abstract, the presenter can prepare for the talk itself by writing it out—not for reading it to the attendees, but simply to rehearse. The following script was used to rehearse the conference paper described in the abstract in Ex. 3.9, with a time allowance of eight minutes.

Ex. 7.1

Differences in ethanol self-selection among inbred mouse strains were first demonstrated by McClearn and Rodgers. In a situation involving a choice between 10% ethanol and water, mice of the C57 strain drink as much as 90% of their daily fluid from the ethanol choice, while mice of the DBA strain almost totally avoid the ethanol solution. Efforts to gain insight into the nature of metabolic factors possibly underlying self-selection have centered around interstrain differences in metabolic capacity, such as those found between high- and low-drinking strains in their ability to clear ethanol and its toxic metabolite, acetaldehyde, from the blood. It has been suggested that strains which avoid ethanol, such as the DBA mice, do so because they learn to avoid the ill effects of accumulated acetaldehyde in their blood.

Less frequently considered has been the possible role of the central nervous system, the site where all alcohols exert their most pronounced effects as narcotics. The recent findings of Schneider et al. indicate that tolerance to and selection of ethanol are positively related, the low-ethanol-selecting DBA strain having a lower tolerance to ethanol challenge than the high-ethanol-selecting C57 strain. A positive relationship between tolerance and self-selection was found more recently by Hillman and Schneider using the alcohol and central nervous system depressant 1,2-propanediol, which, unlike ethanol, is not converted to a toxic metabolite. The C57 and DBA strains are widely separated in their tolerance to and selection of 1,2-propanediol in the same direction as for ethanol.

We have pursued this approach further using the alcohol and central nervous system depressant 1,3-butanediol, a compound of low toxicity that is well suited for investigating self-selection, tolerance, and metabolic relationships in inbred mouse strains.

Male mice of the C57BL/6j and DBA/2j strains were purchased from the Jackson Laboratory in Bar Harbor, Maine, and were about 10 weeks old at the beginning of the experiments.

In the first experiment, 15 mice from each strain were tested for their selection of 10% solutions of 1,2-, 1,3-, 1,4-, and 2,3-butanediol, with distilled water as the alternative choice, for 10 days. Measurements of the amount of fluid consumed from each choice were taken every 24 hours, and an index of selection for each animal was derived by dividing the amount of alcohol solution consumed by the amount of alcohol solution plus water consumed. *First slide.* The first slide is a strain comparison of the 10-day mean selection ratio obtained with each alcohol, and shows that the C57 strain had a significantly higher mean selection ratio than the DBA strain for all of the alcohols except 1,4-butanediol. The widest strain difference in selection occurred with 1,3-butanediol. *Second slide.* This slide shows the mean selection pattern of each strain for 1,3-butanediol over the 10 days of testing. No overlap occurred between the strains on any of the 10 days.

In the second experiment, the open field activity of approximately 15 animals from each strain was measured following an intraperitoneal (i.p.) dose of 1,3-butanediol. The apparatus used consisted of a circular field 14.5 inches in diameter and 8 inches high, with six pie-shaped areas where the animal's movements cut off one of three light beams, resulting in the triggering of relay systems connected to three counters. Thirty minutes following injection, the animal was placed in the apparatus and its movements were recorded for 15 minutes. *Third slide.* This is a comparison of the average strain open field activity at three different doses of 1,3-butanediol. The activity of each animal was expressed as a percentage of the average activity of its own saline control group, represented by the X-bar line on the graph. At the lowest dose tested, the DBA strain was significantly more active than the C57 strain. At the highest dose tested, the DBA strain was slightly less active than the C57 strain. In addition, the rate of drop in activity from the lowest to the highest dose was greater in the DBA strain than in the C57 strain. These results indicate that the DBA strain has a lower tolerance to the overall effects of 1,3-butanediol than the C57 strain, and that a positive relationship exists in these strains between tolerance to and selection of 1,3-butanediol, as found previously with 1,2-propanediol and ethanol.

In the last experiment, whole liver homogenates were assayed to determine liver alcohol dehydrogenase activity in each strain, with 1,3-butanediol and ethanol as substrates. The rate of reduction of NAD was followed at 340 nanometers on a recording spectrophotometer. *Last slide.* The table in this last slide shows the specific activities obtained for liver alcohol dehydrogenase with liver extracts from each strain. The extracts of the high-drinking C57 strain showed higher specific activities than those of the low-drinking DBA strain for both ethanol and 1,3-butanediol. The C57 strain dehydrogenated ethanol at a rate somewhat higher than it did 1,3-butanediol, while the DBA strain dehydrogenated both alcohols at nearly the same rate. Each value shown is the average of two determinations.

The purpose of this investigation was to view self-selection in terms of its relationship to central nervous system sensitivity and tolerance to alcohol and to metabolic capacity. The main advantage of using 1,3-butanediol is that, unlike ethanol, it is not converted to a toxic metabolite, while like ethanol it produces central nervous system depression as well as excitation. The finding that 1,3-butanediol is also differentially selected by high- and low-drinking strains implies that a factor other than toxicity is involved in mediating self-selection. The results of this investigation show that tolerance to and selection of 1,3-butanediol are positively related, as is true for ethanol and 1,2-propanediol. It is also of interest to note that i.p. doses of 1,2-propanediol are less centrally depressive than equimolar i.p. doses of either ethanol or 1,3-butanediol, and that the low-drinking DBA strain consumes 1,2-propanediol in considerably higher amounts than the other two. Therefore, if a common factor is involved in the selection of all of these alcohols, it is possible that tolerance plays a major role.

The strain differences found in the specific activity of liver alcohol dehydrogenase using 1,3-butanediol as substrate are in the same direction and of nearly the same order found with ethanol, and support previous findings by others. However, the correlation of these findings to self-selection and tolerance remains unclear. Recently, Heston et al. showed that mice bred selectively for their differences in ethanol-induced sleep-time had virtually identical liver alcohol and aldehyde dehydrogenase activities. Since a number of factors other than alcohol dehydrogenase activity determine the rate of ethanol metabolism in the intact animal, *in vitro* findings of interstrain differences in metabolic capacity must be confirmed by *in vivo* studies. If both central nervous system sensitivity and metabolic capacity influence self-selection, they may do so independently of one another. In addition, it is possible that the C57 and DBA mice do not have the same mechanism of tolerance to alcohol itself. The finding in this study of a difference between the C57 and DBA strains in their excitability to a low dose of 1,3-butanediol provides a further opportunity to gain a broader understanding of the mechanism of tolerance to alcohol in these strains. *Thank you.*[4]

These nine paragraphs of approximately 1,160 words were rehearsed extensively and timed at slightly under eight minutes. Since the work was completed for an MS thesis, opportunities were plentiful for rehearsing the script with labmates and other department members. The paper follows a straightforward IMRAD style, with the following sequence:

• *Paragraph 1:* Introduces the subject of differential alcohol drinking in laboratory mouse strains and the main causal theory being pursued currently: "metabolic capacity"

- *Paragraph 2:* Provides background to an alternate and insufficiently examined causal theory: neural tolerance
- *Paragraph 3:* States the purpose of the research to be reported presently: to test further the neural tolerance theory with the alcohol 1,3-butanediol
- *Paragraph 4:* Identifies the animal subjects: the mouse strains (DBA, C57), their commercial vendor, and their age
- *Paragraphs 5–7:* Present the methods and results of three experiments to test, respectively: self-selection of butanediols, neural tolerance to 1,3-butanediol, and the rate of metabolism of 1,3-butanediol
- *Paragraphs 8–9:* Discuss the overall significance of the results in the context of prior related work and affirm the possible role of neural tolerance in determining alcohol self-selection differences in these animals

Details, references, and visuals are kept to a minimum in the paper, with the expected emphasis (half the words) on the new experimental work and results. Only four sources are cited, three in the introduction and one in the concluding discussion, while the considerably more detailed journal version published 16 months later cited 33 sources spread throughout the paper.[5] This rehearsal script also makes use of prompts or cues. The first three slides have numerical references ("first slide"), but the fourth is prompted for announcement as the "last slide." The closing "thank you" is both a courtesy and a cue signaling the end of the talk. Finally, there are just four simple visuals (shown below in Figure 7.1). Presentation visuals typically are accompanied by some writing, such as axis labels in graphs or column headings in tables. Electronic slides or frames may project writing in the form of subject or section headings as well as bulleted items. All such writing must be easily and quickly readable by a time-constrained audience. The various forms of writing used in preparing and delivering a presentation must complement the effectiveness of the spoken words to facilitate the listeners' reception and processing of the presented information.

VIEWING

Like a presentation's spoken and written words, the language of visual images must convey its meanings simply and efficiently. Visuals must be easy to follow and absorb under the time limitations. Whether one uses photographic slides or electronic frames, time will allow for just a few carefully selected and simply designed images. The appropriate number of visuals to use will depend on their purpose, type, amount of information, and how readily viewable or readable they may be. One example of a guideline for slides, adaptable to other projected forms, is a measure called a "slaud" (sl).[6] Based on the computer term "baud," the unit value of one slaud is defined as a slide per minute, and the rate recommended is about 0.3 sl. In other words, showing and discussing one slide every 3 minutes—or 3 to 4 slides for a 10-minute talk—allows sufficient time to keep a visual projected so viewers can process the information fully. Increasing this rate poses a risk to keeping an audience's attention on the unfolding story.

Besides their optimal rate of display, effective presentation visuals rely on various elements of design that make them easier to follow and decipher. Such elements function to integrate visuals with the oral text and to facilitate quick viewing and comprehension. Visuals are commonly shown on projected slides or electronic frames. The benefit to an audience of such projected images— whether bulleted lists or graphed experimental data—can be maximized by preparing them with close attention to the following features:

- *Textual connection.* Provide an oral caption that both explains the projected image and connects it smoothly to the scientific narrative. Whereas visuals in written papers are accompanied by descriptive titles and explanatory captions, presenters must give this contextual information orally. The references to the four slides in the sample rehearsal script above illustrate how the transition from spoken text to visuals can be simple, concise, and direct.
- *Prompt identifiers.* Orient readers promptly with key descriptors. These include a figure number, title, labels, and headings (for graphical axes, tabular columns and rows, photographic regions, and so on). A title at the top in larger type, such as 18–24 point, will draw attention more readily to the image's purpose in the paper.
- *Limited information.* Keep slides or frames simple and uncluttered. A frame that is limited to two formulas or curves is more effective than one that is overloaded with data. Tables should have just a few columns and rows of data (about seven or fewer of each). Similarly, wordy and lengthy lists on outline slides make for slower reading than focusing attention on one numbered or bulleted line at a time.
- *Simple design.* Avoid distracting uses of the bells and whistles available in presentational software like PowerPoint and Corel—sounds, border pat-

terns, background colors, and moving elements that do not facilitate quick viewing and processing of the information. These features will only make it harder to focus on the information that matters.

- *Readable typography.* Design typographical elements for easy scanning. The characters on projected images—numbers, letters, symbols—must be highly readable. Use only one typeface: a sans serif (Helvetica, Arial, Geneva) will work well with the short pieces of text typical in presentation visuals, whereas the more designed and higher-contrast serif typefaces (Times Roman, Courier, Palatino) are preferable for the extended reading in articles or books. Restrict use of different character sizes, boldface, or italics to avoid distractions.

- *Appropriate coloration.* Use colors when helpful, but not just for esthetic effect. Contrast and legibility are optimized if colors work well together, such as yellow backgrounds with black lettering or deep blue backgrounds with yellow lettering. Some viewers may be color-blind, so use greens and reds sparingly.

Rehearsing the presentation's text with the visuals will permit better spot-checking for areas that may need refinement, such as the textual transition to each image.

The rehearsal script above refers to four slides that were shown through a projector during the presentation (Figure 7.1). The slides show experimental results comparing two mouse strains—high-drinking C57 and low-drinking DBA—as to their selectivity, neural tolerance, and rate of metabolism of alcohols. The visuals have a reasonable amount of information, simply designed and clearly labeled. The two mouse strains are identified in all the graphs by prominently positioned keys (cross-hatching, geometric shapes) and in the table's column heads. The first graph has just four pairs of bars, revealing that 1,3-butanediol (1,3-BD) has the widest difference in interstrain consumption. The second graph follows up on this information by focusing the narrative on each mouse strain's 10-day consumption of 1,3-BD. The third graph, with only three pairs of bars, shows results for the three doses of 1,3-BD tested for neural tolerance. In all three graphs, the ordinate (y-axis) label is made easier to read by placing it horizontally at the top rather than vertically along the side. The oral text explains each visual, but the ordinate and abscissa labels in the second and third graphs, respectively, could also identify 1,3-BD as the alcohol tested. Finally, the table on the fourth slide is also designed for easy scan-

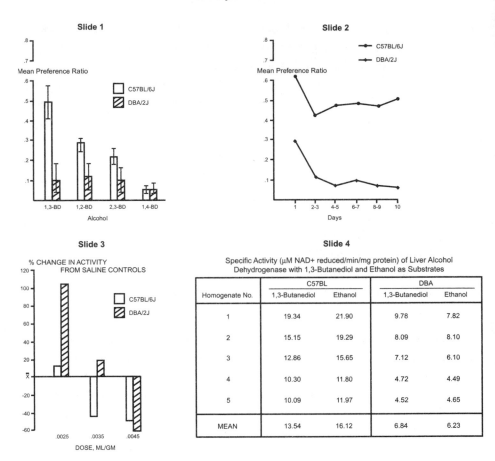

Figure 7.1 Four slides used for an eight-minute conference presentation

ning. Its field has only four data columns. These are sectioned with prominent lines into two pairs of columns, drawing attention to the comparison between the mouse strains (spanner heads) in their metabolism of two alcohols, 1,3-butanediol and ethanol (column heads). Having widely spaced cells also helps make the data easier to grasp. The table's title is sufficiently straightforward, though it could be worded more concisely as "Specific Activity of Liver Alcohol Dehydrogenase," to make it even simpler to read quickly.

Just as in a paper published in a professional periodical, the visual subtext in an orally presented paper frames the story. This is illustrated by the easily followed sequence of slides that accompany the findings reported in our example. Each image's set of data builds toward ultimately supporting a neural sensitivity model to account for alcohol drinking differences. For a talk's visuals to be effective, however, they must deliver information efficiently and in a manner that can be viewed quickly. As important as these visual elements are in aiding the oral delivery of research findings, they are still just that, an aid, and the central focus must remain on the presentation's text itself. Figures and tables are no substitute for adequate preparation and effective oral exposition.

The various considerations that are fundamental for oral presentations—timing, speaking, writing, and viewing—hold importance in their own manner in another type of presentation, the scientific poster. Presenters at professional meetings may opt to communicate their information to attendees via a research poster as a popular alternative to delivering a traditional paper to a seated audience in a conference room.

POSTER PRESENTATIONS

Unlike a formal talk, a poster essentially announces or advertises a research project and, if successful, entices viewers to inspect more closely its purpose, methods, and important outcomes. Posters are exhibited in wide-open spaces, such as hotel ballrooms or corridors, so that attendees may walk about and leisurely view those that interest them. The unique benefit of poster sessions is that the situation is controlled by the *participants,* deciding if, when, and for how long to study the posted information or to talk directly with presenters. Exhibits may last anywhere from a few hours to the entire duration of a meeting. The conference program may list the posters, along with times that presenters will be available for discussion. Their versatility, efficiency, and visual appeal—together with their offer of direct and informal discussion—make posters an important and popular feature of professional meetings.

STYLES OF POSTERS

There are various styles of posters, but they all rely for their success on how information is presented visually (arrangement, typography, color) to draw attention. Once drawn to closer inspection, a viewer can initiate nuts-and-bolts shoptalk that may lead to practical insights for both participants and presenters alike. The style of a poster is determined by several factors: guidelines of meeting coordinators, creativity of presenters, financial resources, means of transport, or even who produces it (the presenter or a graphic artist). Four basic styles of posters are pinup, tabletop, floor-standing, and roll-up. On a pinup poster, each element is mounted on a separate mat board. While pinup posters are easy to transport, they are time-consuming to produce; each mat board must be cut to specification for each element, and all margins between elements must be consistent. A stand-alone tabletop poster may consist of three 2′ × 4′ panels connected at the back to allow folding for transport. Floor-standing posters require more expensive materials and are more difficult to ship or carry to the site. Finally, a roll-up poster is produced on 4′ × 6′ (or × 8′) paper and is laminated. It rolls up for transporting and can be tacked onto a corkboard or other surface. In constructing any type of poster, it is helpful to anticipate not only all the resources and materials that will be needed—mat boards, adhesives, straight-edge, cutting tools, a suitable work space—but how the poster will be packed to move it and what accommodations are provided at the display site. In all these regards, it is also useful to consult with experienced colleagues.

DESIGNING POSTERS

Beyond differences in medium and scale, designing a poster calls for the same considerations as in preparing oral presentation visuals—amount and organization of information, integration of text and graphics, coloration, typography, and host guidelines. The roll-up poster seen here was awarded "outstanding poster" recognition at a national convention of the Association of Women's Health, Obstetric, and Neonatal Nurses (Figure 7.2). The poster reports the process and results of a state-level, multiagency, collaborative effort, led by the Indiana Perinatal Network (IPN), to develop intervention resources for postpartum depression (PPD). A major outcome of the project was a *Consensus Statement on Postpartum Depression* that reviews the scientific literature to provide a comprehensive picture of the condition, treatment options, and intervention resources in the health-care system.[7]

The poster follows several design practices that make it inviting and easy to view. First, the title is written prominently across the top in white letters on a subdued but warm lavender background, and it is positioned next to a picture of an evidently distressed woman whose demeanor fits the subject and elicits a natural curiosity in viewers. The names and affiliations of the poster's authors were made available on business cards.

Second, the layout of the posted items is dynamic, funneling the viewer's eye toward the large graphic in the center that displays the logos of the project partners. The color of the heading on this graphic ("Partners in Healthcare")

Figure 7.2 Roll-up poster that demonstrates the process and results of a collaborative multiagency health-care project to develop a state-level consensus statement on postpartum depression (Courtesy of Joanne Goldbort and Dena Cochran, Union Hospital, Terre Haute, IN)

matches the lavender background, which helps unify all the items. Overall, the use of space and the relative proportions of the components—title, documents, and graphics—are well balanced. The elements designed to attract attention, the title and the logos in the center, are clearly legible from two or three yards away.

Third, the information itself is displayed like a mosaic, providing an overview rather than the minute details. Those who come in for a closer look will see a handful of documents giving more detailed information: a summary of the committee process that produced the Consensus Statement, the group's presentation abstract, and a list of members of IPN's subcommittee on PPD (the three items on the left side of the poster). On the right are the first pages of three other documents: guidelines for PPD care, the PPD Consensus Statement, and a brochure. Full copies of these items, with a list of cited sources, are available at the bottom and side of the display board. Viewers also could put their name and address on a sign-up sheet to receive further information.

When a poster reports on experimental work, the details of IMRAD structure are only sketched, with emphasis on the results and accompanying visuals. As in an oral presentation, only a few key citations are needed. A typical layout for an experiment poster is vertical, with the information arranged in columns under the title and the authors' names. The abstract and the introduction may be placed in the first column, for instance, followed by a second column for methods, a third and fourth for results and visuals, and a fifth for conclusions. If the paper script in Ex. 7.1 was presented as a poster, the results and visuals shown in Figure 7.1 could be highlighted centrally in a large window area like the one with the logos in Figure 7.2.[8] Visuals for an experiment poster, whether graphs or other images, must be unified with the rest of the elements, balanced proportionally, highlighted (contrast, coloration), and easy to scan and decipher. To avoid information overloading, similar strategies as in oral presentation visuals are applicable for ensuring simplicity and quick orientation. The number of visuals, as well as their amount of information (curves, bars, or slices on graphs), should be restrained and balanced with textual elements. Finally, as with other posters, color and typography both in and out of the highlighted areas must be selective and coordinated. Warm colors (red, orange, yellow) may seem more appealing than cool ones (blues and greens), but they can also overpower a poster's message, so balance is the key. Likewise, typographical choices involving character size and typefaces (serif versus sans serif) will affect viewer appeal.

Making scientific posters has become increasingly popular with undergraduate and graduate students, who are invited to present them at some national and local meetings. Such events and competitions teach the value of exchanging scientific knowledge directly. Students may seek guidance for poster preparation from their instructors, campus media resources, and research manuals like the American Chemical Society's style guide.[9]

PREPARING FOR AN AUDIENCE

As one envisions and prepares a scientific presentation, whether a talk or a poster, it is helpful to keep in mind that most of the audience is likely to know less than the presenter about the subject. Only a few participants may have a personal interest or professional experience in your research niche. For a talk, it cannot even be assumed that everyone has read the abstract beforehand. Therefore, for some speakers or situations, it is a good idea to try to establish a rapport with attendees from the first words. Although the most natural and direct opening for a talk simply states its purpose, for instance, an alternative beginning can be used to first engage the listeners' imagination. This could mean using a quote from a widely known authority on the subject, a thought-provoking (rhetorical) question, a striking comparison, or even just an unusual example or bold perspective. Rapport can also be achieved by keeping the discussion straightforward and simple, uncluttered by avoidable technical jargon (as in statistical analysis). Whatever the mix of participants—specialty peers versus relative outsiders—the focus should remain on the fundamental scientific concepts.

For a talk, once an outline, notes (or script), and visuals are prepared, rehearsing will help you to spot and refine any rough areas, adjust vocal force and pitch, check for professional demeanor, manage timing, and generally boost confidence. While a poster cannot be rehearsed in the same sense as a talk, mental preparation for its prospective viewers does nonetheless apply. It helps to realize in advance that most poster viewers typically will stop by for only a minute or two. Therefore, for the session to be of maximum value to the participant in those brief moments of observation and exchange, the presenter must quickly assess and serve a viewer's particular interests promptly, concisely, and articulately. Ultimately, the presenter's demeanor and perceived attitude toward the communicated material become in effect—so far as listeners or viewers are concerned—the work itself. Considering the time and energy asked of attendees, an effective scientific presentation holds attention not only with its content and clarity, but also with the life and force that the presenter breathes into it. It should ideally be an experience that engages the professional imagination of its audience to see the story through to its conclusion.

For a paper presentation, there is a final opportunity to create interest in the subject matter and in the details of the presented work: the question-and-answer period. Although the few minutes devoted to fielding questions may seem like a mere formality to be endured, those moments actually may leave the most lasting impression, for better or worse, of the speaker and the work. Attempts at humor, or casual denigration of the related work of others, may leave listeners with second thoughts about the speaker's motives or trustworthiness rather than having any desirable effect that the speaker may have sought. Such awkward matters aside, questions either from audience members or from a coordinating panel may reveal some quirky details regarding the experimental observations that are not discernable from the more formal presentation itself. A presentation's description of changes in the measured level of motor activity in mice following their injection with ethanol, for instance, may prompt simple questions regarding visual observations: How did the mice appear to behave following such injections? Were they uncoordinated? Did they stagger? Were there any practical limitations or adjustments of the methodology that can be shared only informally, as happenstances of the trade? The personal observations offered by a presenter during those moments of open discussion may, for some listeners, turn out to provide the most valuable scientific insights. It is well, finally, to keep responses to questions as succinct as possible, if for no other reason than to allow maximum audience participation during the available window of time between presentations. These considerations naturally will apply as well, in their own context, to the process of exchange that occurs during poster presentations.

SCIENTIFIC DISSERTATIONS

If I were asked for a single measure of scholarship, a single indicator of dis-
ciplined thinking, and therefore the best single criterion of a good thesis,
I would put forward a plea for simplicity.
—Edwin L. Cooper, "Preparation for Writing the Doctoral Thesis"

THE ROLE OF WRITING IN GRADUATE SCIENTIFIC EDUCATION

While a baccalaureate curriculum affords glimpses of the unique demands
of scientific writing, graduate training serves as a more direct initiation into
the professional discourse of the scientific community. Advanced undergraduates have experiences that interconnect with those of graduate study, such as
doing independent research supported by a professor's grants, presenting a paper or a poster at a conference, or writing a senior honors thesis involving laboratory work. The connection is felt even more deeply in those rare instances
when an undergraduate is listed as a co-author of a journal article.[1] In their
transition from preprofessional to professional writing, graduate students
must demonstrate competence in using the language of science, both spoken
and written, as a rite of passage into their research community. Their research
writing, which typically culminates in either a thesis or a dissertation, is subjected to rigorous scrutiny in its various forms. These include laboratory and

field notes, laboratory reports, course papers, qualifying examinations, a thesis or dissertation proposal, and—after completing the dissertation itself—probably presentations, articles, and grant proposals. This chapter focuses on that unique experience of writing a doctoral dissertation, which attests at once not only to the candidate-writer's scientific knowledge and research ability in a defined area but also to the effective practice of written scientific discourse. Although a dissertation and a master's thesis share an expected rigor in the use of scientific language, a dissertation requires research and writing that has a greater scope and length, with an organization that also lends itself to greater prospects for publication. For career researchers, the dissertation is seen as the ultimate qualifying test of one's training and authority. The importance of a dissertation as both a writing process and a written product must be seen in the broader context of the professional community that cultivates an authoritative and competent use of its tribe's language. How and when does this professional writing evolution begin?

LEARNING SCIENTIFIC WRITING IN GRADUATE SCHOOL

Given the centrality of writing as a qualifying competence for an advanced scientific degree, how do graduate students in the experimental sciences make the transition to becoming professionally rigorous scientific writers? Beyond the personal writing competence that any individual brings to the table, a sense of written and spoken scientific prose is internalized early in a graduate program through various influences:

- hearing classroom lectures;
- receiving critiques on course papers or presentations;
- noticing the variation in prose style among authors of journal articles;
- learning from the models of lucid and fluid prose of important scientific authors;
- attending seminars and conferences to hear how scientific language operates;
- visiting faculty labs to inquire about their work;
- comparing notes with fellow graduate students.

As a doctoral program evolves, a key shaping influence in the student's scientific writing experience comes into play: the dissertation adviser. Once an adviser is found and a dissertation committee is formed, the student's immediate readership becomes clear, and a personal apprenticeship begins. An adviser's pointed critiques of a candidate's writing—whether of lab notes early on or a thesis chapter later—are essential in a successful apprenticeship. When they are ready to begin writing the dissertation, students also have available their own institution's guidelines and numerous published guidebooks.[2] They may also peruse recent dissertations, usually available in a departmental library.

Notwithstanding these diverse resources, the completion of a graduate degree does not guarantee that its recipient will exhibit the most effective style of scientific prose. Medawar states bluntly: "I feel disloyal but dauntlessly truthful in saying that most scientists do *not* know how to write, for insofar as style does betray *l'homme même,* they write as if they hated writing and wanted above all else to have done with it." Such a state of affairs regarding scientists' writing may be rooted in a historical disdain for fussing too much over language in favor of the view that the data speak for themselves. It is also likely the case that graduate students across disciplines would benefit from more formal instruction in the graces of clear, simple, and direct prose to combat common obstacles to readability of scholarly writing. The CBE echoed Medawar's sentiments in its guide for teaching scientific writing to graduate students, the 1986 preface to which cast the problem as follows: "The members of the Council of Biology Editors, like all editors of scientific journals, are acutely aware that many scientists write badly. It is no longer the exception but the rule that scientific writing is heavy, verbose, pretentious, and dull. Considering that the scientists who produce it have received advanced university training, this is little less than shocking. We asked ourselves why these highly educated men and women should express themselves so obscurely, so wordily, and therefore so ineffectually." As a prime contributing factor, the CBE pointed to the paucity of formal instruction in scientific writing, and especially in the writing of effective papers and theses.[3]

Perhaps due to a heightened recognition of the special role of graduate programs in teaching the effective use of professional discourse, graduate courses in scientific writing have grown steadily since the 1980s.[4] There is much room left for improvement, but today there are scientific writing courses under such academic auspices as engineering, basic and health sciences, forestry, environmental science, technical communication, and English. There now appears to be a greater consciousness of the culture of writing in each academic discipline, including the sciences. There are graduate courses and many guidebooks that focus on writing effective papers or grant proposals, skills needed to further one's professional life after graduation. Within this broader acculturation to scientific writing during graduate study, there is the doctoral student's immediate concern—completing a dissertation.

QUALITIES THAT DEFINE A SCIENTIFIC DISSERTATION

Once coursework and qualifying examinations are completed, in consultation with a faculty member heading a dissertation committee, the student proposes an original research project of appropriate scale, duration, and significance. The dissertation project requires several phases: searching the subject's literature; designing the project; collecting data; keeping careful experimental notes; and writing up the work into the book-length dissertation itself. What is a scientific dissertation? What are its basic qualities and expectations? The terms "thesis" and "dissertation" can be used interchangeably (one often hears the expression "doctoral thesis"), but the common distinction followed here is that a thesis is written for a master's degree, a dissertation for a doctorate. Theses and dissertations have much in common as well as significant differences. Both exhibit the degree candidate's disciplined thinking, specialized knowledge, research ability, and, finally, competence in scientific writing. Both test a candidate's ability to function as an independent researcher, though a thesis may be supervised more closely. They also share the IMRAD organizational model. The research and writing for a thesis, however, typically are much shorter in length and time, depending on the specific field, compared with a dissertation.[5] These differences in scale and depth speak to basic differences in sophistication of experimental design and originality of thought. Thesis projects are smaller, simpler, and may even replicate prior research under modified experimental circumstances, and the writing process is intended for completion in one or two semesters. Dissertations are expected to be closer to the originality, significance, and rigor of journal articles, calling for "the same self-discipline, the same hard thinking, and clear, logical, concise writing."[6] In fact, while the progression from chapter to chapter must cohere around a fundamental hypothetical focus, a dissertation's chapters typically are approached as a series of separate experiments intended for potential publication as individual journal articles.

One may argue, as CBE's Edwin Cooper has noted, that the defining qualities of a scientific dissertation are that it

- is an educational tool;
- is the result of individual versus team research;
- may present more than one topic;
- presents a formal statement of hypothesis;
- contains a detailed review of the literature;
- presents all the data obtained in the study;
- offers an extended and argumentative discussion;
- summarizes the results and conclusions;
- lists a comprehensive bibliography.[7]

The first three items are indeed fairly specific to the function of a dissertation, but the other six are qualities shared by journal articles; treatment of these points, like articles themselves, should be kept more restrained and selective. Various decisions face a dissertation writer: How detailed a literature review? How much data to include? How extensive and free-ranging a discussion? These types of decisions are juxtaposed on the overall concerns with format, mechanics, and achieving an authoritative, rigorous, and readable style of scientific prose. Without losing sight of such concerns, let us take a closer look at a dissertation's typical parts.

THE PARTS AND STRUCTURE OF A SCIENTIFIC DISSERTATION

Like books, scientific dissertations (and theses) are organized into chapters and have various items of front and back matter (as well as hard covers). They also follow the IMRAD model for communicating experimental work, especially in biological and chemical disciplines. The general structure is as follows:

- *Front matter:* separate pages for title, copyright, official signatures, dedication, acknowledgments, abstract, table of contents, and lists of tables and figures

- *Text:* chapters that introduce particular experimental activities, describe their methods, present their results, and discuss their implications
- *Back matter:* references and appendixes

Beyond the references in back matter, few dissertations have appendixes, which may include such items as supplemental visuals or reprints of articles published by the candidate while completing the dissertation research or related work. The best way to get a real sense of what scientific dissertations are like is to look at some, especially those related to one's own research interests. Perusing other students' dissertations early in the research process will provide a sense of the road ahead and of certain kinds of textual dynamics, including types of visuals, which may apply to one's own anticipated work. Here we will use an extended example based on both laboratory and field research, submitted in 1998 to the University of Pennsylvania for a PhD in biology, by Michael James Angilletta, Jr. Its title is *Energetics of Growth and Body Size in the Lizard* Sceloporus undulatus: *Implications for Geographic Variation in Life History* (henceforth called *Energetics*). Doctoral students should of course consult the specific format and style manual issued by their own institution.[8] These manuals provide information and guidance in such matters as microfilming, copyrighting, printing, layout of individual pages, spacing, visuals, citation, manuscript style (as found in the *Chicago Manual of Style,* for instance), and depositing the dissertation (including co-submission of electronic copies, on a CD or otherwise).

TRADITIONAL FEATURES OF FRONT MATTER

Whether one is writing a dissertation or just consulting someone else's, the front matter provides important identifying and orienting information. Front-matter pages, totaling 18 in *Energetics,* are numbered in roman numerals, versus the arabic pagination in the main text. The title page, though left unnumbered, is counted as the first page. It contains the same information as the cover, but it adds lines at the bottom for official signatures—adviser, committee members, and department chair—attesting to the work's authenticity and acceptance. This title page has a typical top-down sequence of specific and important items: title, author, the word "dissertation," department, institution, degree, year, and official signatures (a variable number), as shown in Ex. 8.1.

Ex. 8.1

ENERGETICS OF GROWTH AND BODY SIZE IN THE LIZARD
SCELOPORUS UNDULATUS: IMPLICATIONS FOR
GEOGRAPHIC VARIATION IN LIFE HISTORY

Michael James Angilletta, Jr.

A DISSERTATION

in

Biology

Presented to the Faculties of the University of Pennsylvania
in Partial Fulfillment of the Requirements for the Degree of
Doctor of Philosophy

1998

_____

Supervisor of Dissertation

_____

Graduate Group Chairperson

The second page of front matter in *Energetics* (also unnumbered) is for the copyright that claims intellectual ownership, with only three centered items: the word "Copyright," the author's name, and the year. The third page (now numbered "iii") contains the dedication, in this case a 10-word line expressing special gratitude for a spouse's "patience and love" in support of the author's labors. Dedications range from a couple of words to a few lines. Following its title, copyright, and dedication pages, *Energetics* has several traditional items that are longer—the acknowledgments, abstract, table of contents, and lists of visuals.

A dissertation's acknowledgments feature traditionally recognizes the guidance provided by the adviser and dissertation committee members, and thanks other significant supporters. The acknowledgments in *Energetics* (pages iv–v) are organized into six paragraphs that mention, respectively, the names and contributions of the adviser; committee members, as well as faculty who provided laboratory resources; research hosts, assistants, and collaborators, along with commentators on the research and writing; financial supporters; faculty and students in the Biology Department; and family and friends.

The next standard feature of a dissertation's front matter, an informative abstract, varies considerably in detail and length. Centered above the abstract is a repetition of the dissertation's title and author. To be a useful and effective encapsulation of the research, an abstract need not be lengthy and densely packed with detail. The abstract in *Energetics* (pages vi–vii) uses just two paragraphs totaling nine sentences to describe succinctly the experimental work's rationale, findings, and significance.

Ex. 8.2

The fence lizard, *Sceloporus undulatus,* provides a unique opportunity to investigate the causes of phenotypic variation, because it ranges over half of North America and life history traits vary by as much as twofold among populations. *Sceloporus undulatus* exhibits latitudinal patterns of life history that are consistent with those observed in many ectotherms; lizards at low latitudes exhibit fast growth, early maturity, and small body size, relative to lizards at high latitudes. The covariation between latitude and life history suggests that the thermal environment is a major cause of life history variation in *S. undulatus.*

I studied the thermoregulatory behavior of lizards in two populations of *S. undulatus* (New Jersey and South Carolina), and its consequences for the rates of physiological processes related to growth. The thermal sensitivities of metabolizable energy intake and maintenance metabolic rate were quantified over a range of body temperatures experienced by field-active lizards. Despite major differences between the thermal environments of New Jersey (NJ) and South Carolina (SC), lizards in both populations used behavioral thermoregulation to maintain a body temperature that maximized net energy assimilation (= metabolizable energy intake − maintenance metabolism). However, three findings suggest that the annual production budget of a SC lizard is greater than that of a NJ lizard; SC lizards have: 1) a higher metabolizable energy intake at average field body temperature, 2) a lower maintenance metabolic rate at average field body temperature, and 3) a longer daily exposure to preferred body temperature in spring and fall. Thus, SC lizards have a greater potential for growth during the active season than NJ lizards. I argue that latitudinal patterns of growth and body size in *S. undulatus* are caused by variation in the growth potential of lizards and the seasonality of reproduction.

The two concise paragraphs are coherently organized and interconnected, with logically sequenced sentences that follow the IMRAD model of scientific exposition. The first paragraph (three sentences) begins broadly by introducing the subject, "the causes of phenotypic variation," and pointing to the especially suitable characteristics—wide geographic range, substantial life history variation—of the organism to be studied, the fence lizard. The focus then narrows in the second sentence to how particular phenotypic traits (growth rate, body size) in fence lizards co-vary with geographic latitude: "lizards at low latitudes exhibit fast growth, early maturity, and small body size, relative to lizards at high latitudes." The paragraph's third and final sentence states the specific hypothesis to be tested, namely, that "the covariation between latitude and life history suggests that the thermal environment is a major cause of life history variation in *S. undulatus.*"

The abstract's progressive specificity continues in the second paragraph (6 sentences), which begins effectively in two ways: first, it signals a break from the introductory information and draws attention to the author's own work by use of the first person ("I studied . . ."); and second, it provides a transitional link between the two paragraphs—that is, between the key concept in the hypothesis of a *thermal environment* and the study's aim to measure the physiological effect on growth of *thermoregulatory behavior* in two populations (NJ, SC) of fence lizards. Following this statement of purpose, the second sentence moves to the next IMRAD component, methodology, stating what the author actually did: "The thermal sensitivities of metabolizable energy intake and maintenance metabolic rate *were quantified* over a range of body temperatures experienced by field-active lizards." The paragraph's third sentence provides the next IMRAD component, the results: "Despite major differences between the thermal environments of New Jersey (NJ) and South Carolina (SC), *lizards in both populations used behavioral thermoregulation to maintain a body temperature that maximized net energy assimilation* (= metabolizable energy intake − maintenance metabolism)." Finally, the fourth through sixth sentences complete the IMRAD structure with a discussion of the work. They discuss the results' significance ("*three findings suggest . . .*"), set forth a conclusion ("*SC lizards have a greater potential for growth . . .*"), and affirm the study's hypothesis in terms of its findings (again in first person): "I argue that latitudinal patterns of growth and body size in *S. undulatus* are caused by variation in the growth potential of lizards and the seasonality of reproduction." In just nine sentences, tightly constructed and logically ordered in the IMRAD style, the abstract in *Energetics* presents a full and coherent picture of the writer's experimental work.

Following the informative abstract in *Energetics,* two other traditional features of front matter remain: the table of contents and the lists of visuals. The table of contents (pages viii–xiii) simply lists the chapter titles and their subheads, with start pages, while the list of tables (pages xiv–xv) and list of figures (pages xvi–xviii) provide the number, title, and page of all visuals in the text. A look at the table of contents in *Energetics* reveals that the work is divided into six chapters—an introductory chapter, four chapters that report the writer's original research in the laboratory and in the field, and a concluding chapter that discusses the work's theoretical significance. The table listing for each of the four experimental chapters (2 through 5) shows that they are all organized by subheads of the IMRAD style. Here is how the table lists the sections for Chapter 4.

Ex. 8.3

Chapter 4: Variation in Metabolic Rate between Populations of the Lizard
Sceloporus undulatus

Following the table of contents, which lists the titles and subheads of all the chapters, the two lists of visuals provide the titles of all the tables and figures. Just as the reader gets a panoramic sense of the experimental sequence of events from the IMRAD-style listing on the table of contents, perusing the sequence of titles of all the listed visuals begins to form a mental picture of the specific nature of the work and its central elements. For instance, in the titles for the 18 tables and 29 figures listed in *Energetics,* key themes recur that are associated with the work's subject of life history variation in fence lizards—

including energy, temperature, growth, season, and latitude. More than just a formality, the information provided on the table of contents and on the lists of visuals affords an overall yet detailed picture of the dissertation's experimental road ahead.

The highly formalized layout and design elements of the front matter, like those in the main text itself (citations, visuals, subheads, spacing, typography), demand close attention to mechanical details that can try a writer's patience. Still, such stylistic details do serve important functions for maintaining both readability and a disciplined consistency in communicating research information with professional formality and authority. As the dissertation moves beyond these orienting items in its front matter to the main text, the writing process turns to the more critical questions and decisions regarding how best to write up the completed research (experimental and bibliographic).

IMRAD EXPOSITORY STYLE OF CHAPTERS

Because a dissertation represents a first original investigation, it showcases the student's ability to define a problem, select and apply the necessary methodology for addressing that problem, and then present and discuss the results and conclusions clearly and objectively. These competencies can be demonstrated convincingly "only if the student exercises a strong sense of relevance and functional economy, as we have advocated in the writing of a journal article," according to Edwin L. Cooper.[9] *Energetics* has a typical organizational sequence for scientific dissertations of an introductory chapter, several experimental chapters, and a concluding chapter. While this sequence follows a larger IMRAD structure for the dissertation as a whole, the experimental chapters have their own IMRAD structure individually (as shown in Ex. 8.3 for Chapter 4). This expository model parallels that in journal articles, including the brief "Summary" section heading each chapter that is comparable to an article's informative abstract. Throughout the dissertation, its writer faces questions and makes choices regarding various textual considerations from chapter to chapter and across the IMRAD sequence.

INTRODUCTORY CHAPTER

The first chapter, as the initial encounter with the dissertation's subject and aim, will weigh heavily in the reader's mind in evaluating the worthiness of the described scientific endeavor. As the relevant literature is reviewed, the introductory chapter must provide clear, accurate, and sufficiently thorough answers to two basic questions: What is the scientific problem being studied? Why was this particular problem selected? A thorough answer to the first question should also have addressed the second. From the sizable body of gathered sources, mostly journal articles, the student will have to distinguish the good from the bad and then cite only those sources that are of primary importance for constructing a reasoned statement of the problem. Presenting a clear and concise rationale for the investigation—closely circumscribed bibliographically—demonstrates not simply the ability to read sources, but that the student can produce authoritative and critical writing about the subject.

The 16 paragraphs that make up the first chapter in *Energetics* are grouped into four major sections, respectively titled "Summary," "Introduction," "Proximate Determinants of Growth," and "Overview." The Summary and Introduction sections are common to all six chapters. The single-paragraph Summary section in each chapter provides an informative description of the chapter's contents in the same succinct manner as the abstract for the entire dissertation. Since each chapter has its own Introduction section, with its subset of source citations, the first chapter's Introduction section can concisely address and support the overarching conceptual picture of the research area. The five-paragraph Introduction section in Chapter 1 of *Energetics* makes a sequence of points that progressively narrow to the context and purpose of the present study with lizards. Beginning broadly, the section (1) introduces the important roles of body size in life history, (2) describes laboratory and field studies on how temperature affects growth and body size in geographically widespread ectotherms, (3) explains why the lizard is well suited for such studies, (4) identifies the two specific populations of lizards (in NJ and SC) that are the dissertation's subjects, and (5) articulates an overall thesis. The specific rationale and theoretical framework for the student's own research is stated in the general Introduction's concluding sentences:

Ex. 8.4

The greater gain in body mass by SC lizards would provide more energy for future growth or reproduction. Thus, greater annual growth and earlier maturity of SC lizards may be mediated by behavioral and physiological differentiation between SC and NJ lizards, as well as a longer activity season. Further consideration of the behavioral and physiological processes that influence growth would identify additional factors that can contribute to dif-

---

> ferences in annual growth of NJ and SC lizards, and would improve our understanding of the causes of geographic variation in the life history of *S. undulatus.*

The theme of energy availability and its associated behavioral and physiological processes are then focused upon in the next section, which introduces "Proximate Determinants of Growth." Here the important roles of both body temperature and physiological processes (e.g., digestion, excretion, metabolism) are introduced in relation to the key concept of "net energy assimilation" (NEA), the four-paragraph section concluding with the proposition that "both body temperature and physiology of NJ and SC lizards may contribute to greater annual growth of SC lizards."

The fourth and final section of the introductory chapter in *Energetics* is an Overview, with short paragraphs that summarize each of the five remaining chapters, headed by the following paragraph that declares directly and in the first person what the writer does in the coming chapters.

Ex. 8.5

In the remainder of this dissertation, I elucidate some of the proximate mechanisms that influence growth in *Sceloporus undulatus.* I compare the physiology and behavior of lizards in NJ and SC populations of *S. undulatus,* concentrating on the means by which the thermal environments of these two populations limit NEA. I will reveal that multiple mechanisms simultaneously contribute to the difference between the annual growth of NJ and SC lizards.

These clearly and succinctly stated objectives that conclude the introductory chapter provide a smooth transition to reporting the series of investigations that carry them out.

The two sections in Chapter 1 that delineate the research problem and review the relevant literature—"Introduction" and "Proximate Determinants of Growth"—do so economically in just nine paragraphs, directly and closely worded. A review of key factors, theories, and prior work on the selected problem can be thorough without becoming, for instance, a repository of other researchers' findings or bogged down by overloaded sentences, excessive detail, or unrestrained citation. The themes introduced here will, in any case, resurface in the introductions and discussions of subsequent chapters, permitting further elaboration and citation in direct connection with the reported experimental work.

EXPERIMENTAL CHAPTERS

The four middle chapters of *Energetics* (2 through 5) that report the candidate's original experimental work are written in the conventional IMRAD style, as in Ex. 8.3 for Chapter 4, a useful reference point here. In the experimental chapters, each component in the IMRAD sequence poses its own subset of questions for the writer.

Introduction

In introducing one component of the dissertation's broader experimental picture, each chapter will compel the writer to ponder anew: Is the problem taken up in that chapter coherently delineated? Is its connection to the background in the introductory chapter evident? Is the literature review thorough yet sufficiently concise and selective? Is the hypothesis clearly stated?

Chapter 4 in *Energetics* investigates *one* of multiple "proximate mechanisms," behavioral and physiological, that affect fence lizards' growth in the field and that could account for their interpopulation (SC and NJ) differences in growth, namely: temperature-specific differences in metabolic rate, or "annual maintenance metabolism." The hypothetical point is first made in the dissertation's introductory chapter, in the section that delineates "Proximate Determinants of Growth," as follows: "Thermoregulatory behaviors of NJ and SC lizards, and environmental constraints on these behaviors, may contribute to the difference in annual growth between NJ and SC lizards." When this hypothesis is revisited as the subject of Chapter 4, the chapter's own introduction reviews the relevant biological background in greater detail, allowing the candidate to conclude the chapter's literature review by identifying a gap in scientific knowledge addressed in the study and then articulating an anticipated finding:

Ex. 8.6
There is evidence that growth and reproduction of *S. undulatus* is influenced by variation in annual energy assimilation among populations (Grant and Porter 1992; Adolph and Porter 1993), but *there are no data to determine if differences in maintenance metabolism contribute to geographic variation*

> *in the life history of this species.* Because *S. undulatus* in SC are active for
> a longer duration each year (Tinkle and Ballinger 1972; Angilletta, unpub-
> lished data), *annual maintenance metabolism of SC lizards should be*
> *greater than that of NJ lizards, unless the temperature-specific rate of me-*
> *tabolism is lower for SC lizards.*

Just as the introductory chapter ends with overall objectives, Chapter 4's introduction ends with a direct and simple declaration (less detailed than the
Summary) of the study's objective, hypothesis, methodology, results, and conclusion:

> Ex. 8.7
> The purpose of this study was to quantify the thermal sensitivity of meta-
> bolic rate for *S. undulatus* from NJ and SC populations. I compared meta-
> bolic rates in three seasons, at temperatures that are experienced by lizards
> in both populations, *to test the hypothesis that SC lizards have lower meta-*
> *bolic rates than NJ lizards.* I demonstrated that lizards from SC do have
> lower temperature-specific metabolic rates than those from NJ, and that this
> difference results in higher daily maintenance costs for NJ lizards. There-
> fore, rates of energy expenditure, as well as rates of energy assimilation,
> contribute to the differences in growth and reproduction in these popula-
> tions.

The introduction to each experimental chapter in the dissertation provides a
context and rationale for the reported research, assisting readers to readily interconnect the lines of thought both within the chapter and to the dissertation's
broader picture. In Chapter 4, we must see the connection between wanting to
"quantify the thermal sensitivity of metabolic rate" (Ex. 8.7) and the larger
aim—announced in Chapter 1's Introduction section—to "improve our understanding of the causes of geographic variation in the life history of *S. undu-*
latus" (Ex. 8.4).

Ultimately, the dissertation's writer must weave a visible thread through all
the chapter introductions and discussions (and the concluding chapter on theoretical implications) that connects back to the thesis first announced in the
front matter, which in *Energetics* we see in its abstract's final sentence (Ex.
8.2): "I argue that latitudinal patterns of growth and body size in *S. undulatus*

are caused by variation in the growth potential of lizards and the seasonality of reproduction." For the scientific narrative to maintain cohesiveness and continuity, there is a necessary degree of reaffirmation of key themes (energy, growth, seasonality) through strategic repetition, in one form or another, across the dissertation.

Methods and Materials

The second IMRAD component of a dissertation's experimental chapters—a description of precisely what was done and how it was done—also has its expository demands. A clear, unequivocal, and suitably detailed picture must be given of the following experimental elements.

- *materials:* commercial and noncommercial products; supplier, lot number, grade, packaging, and expiration date; chemical names, formulas, and properties
- *organisms:* supplier or geographic origin, sampling, number, age, sex, size, housing, feeding, acclimation, seasonal features, other relevant information
- *design:* type or name of experimental design, dependent and independent variables, controls, subject groupings
- *setting:* laboratory or field environment—temperature, humidity, lighting, air quality, pressure, fluctuations during the experiment, and so on
- *equipment:* instruments used, with sketches of any unfamiliar, modified, or innovative features; manufacturers, models, and catalog numbers
- *procedure:* nature, order, and timing of experimental steps and activities; types, names, or modifications of techniques
- *measures and analyses:* instrumental parameters, calibrations, measurement conditions, statistical applications, conversion factors, assumptions, and limitations in accuracy[10]

Naturally, this is where a well-kept laboratory notebook is of great service. The question for the candidate-writer now becomes: How much (or how little) detail is it necessary to extract from the experimental records to describe accurately and completely the methods, materials, and instruments used in the laboratory or in the field? Moreover, how much should the candidate dwell on— or attempt to excuse—all the various methodological limitations, such as errors of omission, problematic techniques, or deficient samples? While a sound dissertation will keep such excuses to a minimum, Cooper says that "some show of humility and an awareness of the fallibility of human endeavor will not be construed by an intelligent reader as a significant weakness."[11] The emphasis is best placed, however, on the positive and successful outcomes of the research.

In our extended example from *Energetics,* Chapter 4's Methods section describes the means used for measuring the fence lizard's metabolic rate in just seven paragraphs that are organized into four subsections: "Animal Collection and Care," "Measurement of Metabolic Rate, "Data Analysis," and "Regression Models." In the study reported in Chapter 4, "the experimental design involved repeated measures of the metabolic rates of individuals." In the single-paragraph "Animal Collection and Care" section, we learn the more specific origin of the studied lizards—Lebanon State Forest in Burlington County, NJ, and the Savannah River Site in Aiken County, SC—as well as their sampling, number, sex, age class, transportation, housing, feeding, and incubation.

Next, the two-paragraph subsection "Measurement of Metabolic Rate" describes the instruments and procedures used to measure the lizards' metabolic rate. Names, models, and manufacturers of instruments—a flow-through respiratory system with programmable incubator, CO_2 and O_2 analyzers, mass-flow equipment, a cloacal thermometer, and a data acquisition program—are specified along with their settings, calibration, and measurement and recording processes. The following excerpt begins with an overall sketch that develops into a more detailed picture of what was done.

Ex. 8.8

Metabolic rates of lizards were measured with a flow-through respirometry system (TR-3, Sable Systems International, Henderson, NV). Sixteen respirometry chambers, each 120 ml, were contained in a programmable incubator (Model 818, Precision Scientific, Chicago, IL). An opening in the incubator, 5 cm in diameter, was used for incoming and outgoing tubing. Incoming air was scrubbed of H_2O and CO_2, and pushed at 150 ml min^{-1} through 20 m of copper tubing submerged in 38 L of water that was at equilibrium with the incubator temperature. . . . Outgoing air was scrubbed of water and entered a mass-flow meter (v1.0, Sable Systems International, Henderson, NV), a CO_2 analyzer (Model LI-6251, LI-COR, Inc., Lincoln, NE) and an Oxygen analyzer (Model FC-1, Sable Systems International, Henderson, NV). . . . Prior to the study, the mass-flow system was calibrated using a mass-flow controller valve (Sidetrak,™ Sierra Instruments, Inc., Monterey, CA) connected to a mass flow controller electronics unit (v1.0, Sable Systems International, Henderson, NV).

Further details regarding the measurement of each lizard's metabolic rate provide the precise measurement protocol, including intervals, number of samples and recordings, weighing of animals, photoperiodicity, temperature settings, and control measures. The procedure for key measurements—of CO_2 and O_2 levels—is described closely:

Ex. 8.9

From 1200 h to 0800 h the following day, the production of CO_2 and the consumption of O_2 were measured for a period of 2 min every 2.5 h, resulting in 8 recordings for each individual. During each 2 min period, concentrations of O_2 and CO_2 in the chamber and the flow rate through the chamber were recorded each second by the data acquisition program DAC (Sable Systems International, Henderson, NV).

Since a dissertation's description of methods generally is expected to be lengthy and fully detailed, the writer who provides a brief overview before moving on to the finer points (as in Ex. 8.8) will have appreciative readers. Even in the thick of the methodological narrative, the candidate may spare certain kinds of details, such as every tried or failed procedure in the course of deriving reliable data, inconsequential anomalies, or standard and widely known laboratory materials and practices. On the other hand, special attention to detail—including visual representation—is appropriate for describing significant modifications or innovations in experimental practice or equipment.

The final two subsections of Chapter 4's Methods, "Data Analysis" and "Regression Models," describe how the data recorded for each lizard on metabolic rates (CO_2 and O_2 concentrations) were used to quantify and anticipate interpopulation effects or geographic variation (the dissertation's theme). The three-paragraph "Data Analysis" subsection explains the procedures used to calculate metabolic rate, to express it as "energy expenditure," and then to compare its intra- and interpopulation effects (using the Statistica for Windows software program for analysis of covariance, or ANCOVA, and "Tukey's honest significant difference test"). While data visuals are most commonly used in a study's results section, the "Data Analysis" section does contain a small table showing the fence lizard's averaged body masses used for measuring metabolic rate (Table 8.1).

Table 8.1 Mean body masses (g) of lizards collected from two populations of *Sceloporus undulatus* in three seasons for measurement of metabolic rate. Standard errors are given in parentheses.

Season	New Jersey	South Carolina
Spring	4.2 (0.8)	4.2 (0.6)
Summer	6.7 (0.8)	3.9 (0.8)
Fall	4.3 (0.9)	4.0 (1.0)

Courtesy of Michael J. Angilletta, Jr., Indiana State University

As with all other visuals in *Energetics,* there is a parenthetical reference to the table, but the table itself is appended to the back of the chapter. Finally, the single-paragraph subsection "Regression Models" describes how the candidate "constructed multiple linear regression models to predict maintenance metabolic rates in each season from body mass, temperature, and time period," including the models' assumptions regarding those variables. The standard statistical terms and procedures—multiple linear regression, ANCOVA, Tukey's test—are identified only by name. If the details that describe procedure, measurement, and analysis are presented fully and unambiguously, readers will be led actively toward wanting to know the experimental outcomes, rather than being left struggling to overcome the inertia of obfuscation. The presentation of results, our next concern, is a dissertation's core content.

Results

Just as order and logic are essential for a coherent narration of a study's methods, a clear presentation of the results means that readers should readily be able to see their connection to both the just-described methods and the study's purposes. This third component of IMRAD poses various questions: How *much* data should be reported and in what order? What type and number of visuals are needed? Does the wording accurately narrate and precisely quantify what occurred? Is sufficient statistical information included for readers to fully understand the results and to assess their reliability and significance? Will readers see the connection of the results to the study's hypothesis? In our extended example of Chapter 4 in *Energetics,* its Introduction already announced what the results show (Ex. 8.7)—"that lizards from SC do have lower temperature-specific metabolic rates than those from NJ, and that this difference results in higher daily maintenance costs for NJ lizards." Now the candidate must organize the Results section so that readers can smoothly follow how the methodology yielded confirmatory data.

The Results subsection in Chapter 4 is organized into six brief paragraphs, each of which consistently first states and then elaborates a particular experimental outcome, respectively headed with the following six simple and declarative topical sentences.

- Respiratory exchange was dependent on temperature.
- The mean metabolic rates of NJ lizards were higher than those of SC lizards in two of three seasons.
- In all seasons, metabolic rate was insensitive to temperature over part of the temperature range that was examined.
- A consistent diel cycle in metabolic rate was discovered in all seasons.
- Metabolic rate varies seasonally in both populations, but the direction of change was different for NJ and SC lizards.
- Regression models of metabolic rate that were used to estimate daily maintenance metabolism were highly significant, and explained 30%–75% of the variation in metabolic rate.

The order of these points parallels the earlier order in the Methods section of procedures described, respectively, to measure the lizards' interpopulation metabolic rates, to reveal variance with temperature and seasonality, and to model differences in "maintenance metabolism." As each finding is explained, more specific patterns of effects (thermal, seasonal, temporal) are noted and compared between the two lizard populations (NJ, SC).

It is also evident from the number of references to visuals—five tables and four figures appended to the chapter—that the Results component of IMRAD relies most heavily on showing rather than just telling the scientific story. Authors naturally have to make decisions about which data must be presented visually, how many visuals will be necessary, and in what forms to present them. In addition, like other prescribed elements, a dissertation's tables and figures must be formatted and placed according to local guidelines, with any discretionary options nonetheless requiring legibility, completeness, consistency, and an awareness of conventional leeway.

As in Chapter 4, the Results section may be IMRAD's shortest textual component—due largely to the condensed presentation of data in visuals—but it is the section that demands the utmost clarity, simplicity, and directness. Care should be taken, for instance, to avoid unnecessary content that clutters the narrative and detracts from the candidate's own scientific contribution and authority, such as tangential discussions that may be interesting but not directly relevant to the present work, or recapitulation of the related results of others to affirm one's own results. Even as to sentence style itself, readers will appreciate being assisted with wording that is consistent, symmetrical, and even monotonously repetitive. A guidebook from the APA on writing dissertations advises: "Decide on a particular sentence structure that most clearly presents the results of a particular type, and stick with that structure for all the results that are similar."[12] Results that are reported fully and in a readily followed manner will by valued by readers who return to these pages periodically as they attempt to make sense of the points raised in the discussion.

Discussion

The final IMRAD component, discussion of the results, calls for a critical look at the study that interprets and assesses the reported findings, pointing up their significance, limitations, and implications in the context of both the hypothesis and current knowledge in the field cited from the reviewed literature. Conclusions must be strictly rooted in the candidate's own findings, and connected to the subject's current theoretical paradigms. In Chapter 4 of *Energetics,* the reader has already been apprised of the study's overall conclusion in the chapter Introduction's final paragraph: "Therefore, rates of energy expenditure, as well as rates of energy assimilation, contribute to the differences in growth and reproduction in these populations." A study's discussion section also requires, like the other IMRAD components, the clearest organization and most logical order of points.

The 14-paragraph Discussion section in Chapter 4, by far the chapter's longest, is divided into four subsections: "Thermal Sensitivity," "Diel Variation," "Seasonal Acclimatization," and "Variation between Populations." Each of these subsections begins by reiterating a particular observation in the candidate's study that is at once compared with other researchers' findings on the same point. The subsection that discusses thermal sensitivity, for instance, begins with a confirmatory observation (italics added): "Metabolic plateaus, temperature ranges over which metabolic rate is insensitive, *are common* in *S. undulatus* and other species of reptiles (Waldschmidt et al. 1987)." However, the discussion will soon progress to points where the candidate's own findings contrast with those reported by others. At these critical junctures, beyond noting such differences in results and offering possible explanations for them, it is also appropriate to suggest future research directions to address the issue more deeply. The following two sentences from the Discussion subsections on thermal sensitivity and seasonal acclimatization, respectively, suggest such future directions with support from both the literature and an earlier chapter's findings.

Ex. 8.10

1. Careful scrutiny should be given to adaptive hypotheses for plateaus of temperature-independent metabolism, particularly because populations of *S. undulatus* differ in their thermal sensitivity of metabolic rate, but not in their thermal sensitivity of other physiological processes (Crowley 1985; Chapter 3).
2. Clearly, acclimatization of metabolic rate is not straightforward and studies should be designed to examine hypotheses that incorporate multiple causality at a biochemical level (Clarke 1993).

Forward-looking perspectives like these, and any other critical pronouncements in the discussion, must be based on and supportable by the candidate's own findings reported in the preceding Results section. Where significant, the discussion should also address the study's limitations and their specific implications, as in these concluding sentences of the "Thermal Sensitivity" discussion.

Ex. 8.11

Importantly, metabolic rates reported here and in Zannoni (1997) are not equivalent to standard metabolic rate (SMR), because measurements were made during periods of photophase, as well as scotophase. Also, neither study selectively reported the minimum rates of metabolism observed, as in many studies of SMR (e.g., Feder and Feder 1981; Tsuji 1988a). Therefore, metabolic rates reported here are expected to be slightly higher than SMR (Niewiarowski and Waldschmidt 1992).

In sentences like those in Ex 8.11 that interweave references to the literature with the candidate's present work, the reader must be able to distinguish which findings or conclusions are whose. One unequivocal marker, used liberally in

Energetics, is a first-person reference like "in my study" or "I found," as in the second of these two sentences (with emphasis added) from the "Seasonal Acclimatization" subsection:

> Ex. 8.12
> Although organisms may undergo reduced maintenance metabolism during periods of energy limitation, faster growth during the periods of energy surplus would be expected to increase energy metabolism (Wieser 1994, 1998). Therefore, *I interpret* the seasonal changes in metabolic rate observed in NJ lizards as the product of increased rates of physiological processes, such as growth.

Critiques and assessments like those shown in this example, which are at the heart of a discussion section, will showcase the breadth and depth of a candidate's scientific knowledge, logical reasoning, and overall professional competence. It cannot be overemphasized that no small aspect of this test of authority and professionalism is the level of rigor in the candidate's practice of critical thinking as demonstrated in the language of the tribe.

The closing thoughts of a study's discussion should return to its hypothesis and reconnect with the dissertation's overarching theme. Readers will appreciate being reminded of both as the candidate sums up the work's significance. The final paragraph in Chapter 4 of *Energetics* begins with a simple declaration of the study's broad upshot (with cited support), followed immediately by a reaffirmation of the study's thesis in the context of the relation between growth and energy dynamics:

> Ex. 8.13
> My study highlights the importance of considering multiple causality of ecological phenomena (Quinn and Dunham 1983). The greater annual growth of SC lizards may be caused by both a higher rate of energy assimilation and a lower rate of maintenance metabolism.

The remainder of the closing paragraph speaks to the importance of developing a more complete picture of the bioenergetics of growth in the lizards studied by synthesizing laboratory findings (like Chapter 4's) with those in the field. This concluding sentence returns to the dissertation's theme of multiple causation in life history variation:

Ex. 8.14

Undoubtedly, a genuine understanding of the causes of geographic variation in the life history of *S. undulatus* will not be achieved by formulating simple causal hypotheses, but will compel an integrative approach designed to tease apart the relative contribution of multiple mechanisms.

The discussion section as a whole essentially is the place for the writer to demonstrate not only critical competence in interpreting and assessing the findings, but also the professional authority necessary to argue for their significance convincingly.

An experimental chapter as a whole, as illustrated here with Chapter 4, demands close attention to a range of expository practices, including coherent and symmetrical ordering of points within and across IMRAD sections; emphasis of key findings and ideas through repetition; judicious citation from the literature; suitable choice and design of visuals; interpretations, speculations, and conclusions that are grounded in the reported findings; and logical progression of thought across sentences, paragraphs, and sections. Using clear, simple, concise, and consistent wording will only buttress the effectiveness of these practices. Once the experimental chapters are completed, the candidate will have to overcome the inertia of feeling, prematurely, that there is nothing left to add, as the dissertation's final chapter is by no means just a mere formality.

FINAL CHAPTER

When it comes to discussing the ultimate meaning and value of the completed dissertation, the candidate, standing alone and most vulnerably before committee members who have mentored the project, is expected now to demonstrate to them lucidly and authoritatively the high level of scientific knowledge and thought that merit full-fledged membership in the professional research community. Following the earlier discussions in experimental chapters focused on the individual studies, the final chapter concludes the dissertation with a more global and comprehensive analysis of the entire project. It launches into deeper theoretical waters, where bold or creative candidates can take analytical risks associated with critiquing established paradigms or arguing for new ones. Given the dissertation's educational value as an initiating demonstration of professional competence, so long as the rules of logic and inference are followed and the focus remains on the topic at hand, Cooper says, "the student can be granted considerable license in speculative thought."[13]

The sixth and final chapter in *Energetics* begins by pointing out that, although likely biological determinants of life history traits in fence lizards have been identified—including phenotypic plasticity and genetic divergence—"the ultimate mechanisms underlying geographic variation in the life history of *S. undulatus* remain obscure." The theoretical focus is evident from this short paragraph that concludes the chapter's Introduction by announcing its aims, stating its conclusion, and looking ahead:

Ex. 8.15

This chapter has three main goals: 1) to review proximate and ultimate models of life history, with respect to latitudinal patterns of life history in *S. undulatus,* 2) to evaluate the plausibility of a recent model of life history evolution (Berrigan and Charnov 1994), and 3) to introduce an alternative model that incorporates the effects of seasonality on growth and maturation. We conclude that simple adaptive hypotheses cannot provide a general explanation for the latitudinal patterns of life history in ectotherms. Future theoretical work should focus on one or a few species, where the proximate mechanisms for variation and growth and age at maturity are well known.

Between the chapter's Introduction and concluding Discussion, there are three major sections, respectively titled "Competing Hypotheses," "Evaluating the von Bertalanffy Hypothesis," and "Seasonality and Life History." The subsections under these headings review selected theoretical models used to account for patterns in life history variation, pointing out shortcomings that render them inadequate for the task and anticipating the need for a new model supported by the candidate's own findings. Various statistical formulations and graphics are used to develop and support the chapter's theoretical critiques as well as the alternate model proposed (goal 3 in Ex. 8.15). The innovative aspect of the newly proposed model is introduced in the context of a prior finding (emphasis added):

Ex. 8.16

Adolph and Porter (1996) showed that seasonality and thermal constraint on activity are sufficient to generate a phenotype that exhibits slow growth, delayed maturity, and large body size in *S. undulatus.* The model developed here (referred to hereafter as *the seasonality model*) is the first attempt to incorporate this idea into an evolutionary model of life history.

The discussion then moves on to describe the assumptions of the proposed "seasonality model" as well as the inferences and predictions that specific applications of the model make possible. The dissertation's closing paragraph emphasizes the proposed analytical model's "preliminary attempt to incorporate the consequences of seasonality into life history theory" and suggests directions for further study.

Ex. 8.17

Future work should focus on two major areas: 1) modeling growth as an allocation process rather than a fixed trajectory (Bernardo 1993), and 2) incorporating multiple causal mechanisms that can produce variation in growth rate (e.g., behavioral, physiological, environmental).

The scientific complexity and sophistication of its arguments notwithstanding, the final chapter of *Energetics* sets forth an original and coherently constructed argument that contributes to the theoretical and ultimately the practical understanding of its narrowly circumscribed subject. This should be the culmination of any successful dissertation.

David Garson summarizes six elements of reasoning that are of paramount importance for a scientific work's concluding discussion: "The 'claim' is the debatable assertion found in hypotheses. The 'grounds' are the evidence supporting the claim. The 'warrant' explains how the grounds support the claim. The 'rebuttal' lists the conditions under which the warrant and grounds support the claim. The 'modality' indicates the level of confidence or certainty the researcher has in the claim. Finally, the 'backing' sets forth the 'givens'—the assumptions on which the whole argument rests."[14] The concluding chapter in *Energetics* effectively includes all of these elements in some recognizable form. These are the features of rigorous scientific thinking that a candidate must showcase, both in the dissertation and in its oral defense. The focus here has been on expository features of a dissertation, without delving into such accompanying and important matters as selecting an adviser, focusing the research problem, preparing and defending the proposal, or working through draft stages. Moreover, although the extended example is a dissertation in biology, the same basic concepts of expository development also apply in some form to dissertations in other scientific disciplines, such as chemistry, physics, and the health sciences. Finally, beyond an intellectualized sense of a dissertation provided here, doctoral candidates must realize that only holding a sample dissertation in one's hands and examining its text will provide a full sense of the product, and that only experiencing such writing will ultimately teach the process.

AFTER THE DISSERTATION

Once the dissertation is completed, successfully defended, and deposited with the graduate school, its writer can consider a further professional option—publishing it, either in book form or, more commonly, as articles. Chapter 4 of *Energetics,* for instance, was published under a slightly modified title in the journal *Physiological and Biochemical Zoology* (see Chapter 9).[15] Dissertation chapters that are written up as discrete studies, like those in *Energetics,* readily lend themselves to being transformed into articles. Since the educational nature of a dissertation requires that it contain more information than would be necessary for publication purposes, its writer will have to reexamine it to condense, delete, or modify content, including visuals, for a wider professional audience. Suggestions for revision may come from the faculty mentors, and other strategies for revision will become apparent from examining sample articles in journals being considered. Once the graduate decides on a particular journal for submission, that publisher's own editorial guidelines includ ing those for format and length—will come into play. New graduates who venture into publication waters should be mindful of the fact that there is a high rejection rate for submissions, especially among the leading periodicals in a particular research niche. A rejection and reviewer critiques, however, may speak less for the overall value of a particular manuscript's contribution than about competition from a large submission volume. Rather than being discouraged, therefore, one should immediately consider resubmission to alternate periodicals that publish on that subject. Preparing one's research for publication as a journal article brings its own professional demands, conventions, and rewards.

SCIENTIFIC JOURNAL ARTICLES

Scientific papers form the scaffolding of science.
—Michael Katz, *Elements of the Scientific Paper*

PROFESSIONAL IMPORTANCE OF JOURNAL ARTICLES IN THE SCIENCES

Given its social and collaborative nature, scientific inquiry is vitally dependent on publication. Graduate students, especially doctoral candidates, are encouraged to publish even during the course of their studies. The productivity
of scientists seeking academic or research positions is gauged largely by their
publication potential or history. When a researcher has made a significant advance—yielding some new result, technique, or concept that engenders new
insights—disseminating it to interested audiences is not merely an option but
a professional obligation. Bentley Glass has written: "Both the international
scope of scientific activity and the cumulative nature of scientific knowledge
lay upon the individual scientist an overwhelming debt to his colleagues and
his forerunners. The least he can do in return, unless he is an ingrate, is freely
to make his own contributions a part of the swelling flood of scientific information available to all the world."[1] Aside from the potential benefits (or risks)
of any scientific advance, just the billions of federal dollars granted annually
to support research would speak loudly for disclosure. Given that science is a

human endeavor, professional jealousy and self-serving secrecy are certainly not alien to today's competitive culture of scientific activity. In our complex world in which science is conducted, the Baconian ideal to freely share scientific knowledge may be subject to conflicting interests—personal, professional, economic, political—acting to thwart that ethical duty. Nonetheless, the scientific community is steadfast in its longstanding tradition and standard expectation that researchers will publish what they learn and do so in a form that is fully recognized as professionally valid. Depending on the particular circumstances or magnitude of a particular scientific finding, preliminary forms of publication—letters, press releases, conference talks, or brief communications in journals—may precede a full-fledged version. Such preliminary announcements were used by Watson and Crick, for instance, in their famed race with other researchers to elucidate the structure of DNA in the 1950s. Beyond such initial statements, the scientific community uses a standard form of publication: the journal article. In due course, the complete details of Watson and Crick's DNA work appeared in such highly regarded journals as *Nature* and *Proceedings of the Royal Society.* As a historical marker, the Royal Society of London held true to its Baconian principles when it began publishing in the 1660s a pioneer scientific journal, its *Philosophical Transactions.*

What is the professional standard of validity that is satisfied by publication in a scholarly journal or in other readily accessible forms of dissemination? One CBE editor supports the following definition for "primary" publication: "An acceptable primary scientific publication must be the first disclosure containing sufficient information to enable peers (1) to assess observations, (2) to repeat experiments, and (3) to evaluate intellectual processes; moreover, it must be susceptible to sensory perception, essentially permanent, available to the scientific community without restriction, and available for regular screening by one or more of the major recognized secondary services (e.g., currently, Biological Abstracts, Chemical Abstracts, Index Medicus, Excerpta Medica, Bibliography of Agriculture, etc., in the United States and similar services in other countries)."[2]

This definition covers publication in visual and nonvisual (e.g., audio) forms, so long as the criteria are met for complete and unrestricted disclosure. As to journal articles in particular, professional peers of the authors must be able to follow, assess, and replicate the presented experimental findings to test their reliability and validity. Even before publication, peer reviewers of a submitted manuscript customarily will evaluate its worthiness as a prospective contribution. This open, complete, and rigorous scrutiny of their content is what makes journal articles such a highly valued form of scientific publication. What exactly is a journal article? How do an article's content and structure work together to effectively transmit what a researcher has done, observed, and concluded?

SCIENTIFIC JOURNALS AND THEIR ARTICLES

Journals in the sciences contain various kinds of items besides articles that report original experimental work, a fact that can bewilder inexperienced readers like advanced undergraduates or early graduate students who are required to cite scholarly sources. Among the types of informational items published in scientific journals besides research articles are news reports, editorials, columns, letters, and book reviews. Journals also have other unique features or "departments" that they publish regularly. Although scientists will readily differentiate among these types and their respective aims, undergraduate students must be taught to distinguish their technical purpose and bibliographic value. Students also need to know that journals vary in their degree of specialization, with some publishing articles on a broader subject range, such as *Science* or *Nature,* and others in narrower specialties or subspecialties, such as *Astroparticle Physics* or *Journal of Electroanalytical Chemistry.*

Whatever the subject domain or scope of a journal, research articles take one of the following typical forms: empirical or experimental, methodological, theoretical, review, and case study. Although there is no established formal typology of scientific articles, these forms may be differentiated in basic content and purpose as follows:

- *experimental:* reports original laboratory or field studies, typically organized by the IMRAD model;
- *review:* synthesizes previously published work to evaluate the state of current knowledge in a defined area, identifying gaps and suggesting future directions;
- *theoretical:* draws on available work, including empirical studies, comparing consistencies and contradictions of alternative theoretical constructs— whether verbal, graphical, or mathematical—to support an existing theory or develop a new one;[3]

- *methodological:* presents modified or new methodologies, such as in laboratory techniques or data analysis tools, permitting practical comparisons with existing approaches in particular research areas or problems;
- *case study:* describes and analyzes quantitative or qualitative information obtained from studying individuals or organizational settings to demonstrate a problem (such as a medical condition or an occupational hazard) or a need for new solutions and theories.

The focus here is on the conventional parts of a scientific article, especially the experimental article and its typical IMRAD structure. As a genre, experimental articles proceed inductively by describing a series of laboratory or field events that lead to a broader statement about natural phenomena. The typical structure of scientific papers, beyond serving to report information in a formalized fashion, is an idealization of scientific inquiry—a simplified progression from experimental design to collection and presentation of results to conclusions about the natural world. The overall structure includes a range of features that allows articles to communicate their content with consistency and maximum readability.

THE FEATURES OF A SCIENTIFIC JOURNAL ARTICLE

Scientific journal articles have conventional components that their writers include, as well as additional in-house features of design and layout determined by publishers. Besides the perennial concerns with effective use of scientific language, authors of articles should attend closely to the following aspects of their manuscript:

- Title
- Author byline
- Abstract
- Acknowledgments
- Textual organization
- Visuals
- References
- Ethics

Once a decision has been made about which periodical the paper will be submitted to, the prospective author(s) should consult the manuscript guidelines for that publication. Also helpful are the general guidelines for preparing manuscripts available in manuals from such organizations as the American Chemical Society (ACS), the American Psychological Association (APA), and the Council of Science Editors (CSE).

WORDING THE TITLE

The title page for an article in a journal typically includes the full title, authors' names and affiliations, an abstract of the paper, and a set of keywords to be used in indexing the article; one such page, from an article in the *Journal of Nutrition,* is shown in Figure 9.1. Deciding on the wording for an article's title is no mere formality or simple matter. The language used in an effective title should be precise and informative yet as concise and unambiguous as possible. The title will serve as a guide or "mini-abstract" not only to those who hold the article in hand but also to researchers sifting through abstracting or indexing databases. For titles to be maximally clear, informative, readable, and readily indexed, authors should do the following:

- select concrete, specific, and precise language to capture the paper's content;
- use unambiguous syntax;
- keep the title as brief as possible, without unnecessary words;
- consider using a subtitle when brevity is difficult to achieve;
- spell out words, avoiding abbreviations, formulas, and symbols.

As one example of how to word a title, the *Journal of Nutrition* provides instructions for authors that read simply: "Include a title which is a declarative statement of key findings and which includes the species studied."[4] The title used for our example in Figure 9.1, "Liver Fat and Plasma Ethanol Are Sharply Lower in Rats Fed Ethanol in Conjunction with High Carbohydrate Compared with High Fat Diets," does indeed satisfy those requirements. The title is helpful as a mini-abstract not only in its comprehensive use of keywords, but in providing the study's results—that is, with the high-carbohydrate diet, liver fat and plasma ethanol levels "are sharply lower" (use of the generalized "are" instead of "were" is still understood for findings in the specific case). Were the instructions more flexible—for example, not requiring a declarative form—such a title also could be effectively worded as follows: "Liver Fat and Plasma Ethanol in Rats Fed Ethanol: Sharp Reduction of Both

Liver fat and plasma ethanol are sharply lower in rats fed ethanol in conjunction with high carbohydrate compared with high fat diets[1]

Hans Fisher,[2] Alycia Halladay,[*] Nagarani Ramasubramaniam, James C. Petrucci
Dennis Dagounis, Anna Sekowski, Joseph Martin[†] and George Wagner[*]

Department of Nutritional Sciences,[*] Department of Psychology[†] and Department of Biology,
Rutgers University, New Brunswick, NJ 08901-0231

ABSTRACT The effects of high fat and high carbohydrate diets on alcohol metabolism were studied on blood alcohol and liver fat concentration. In Experiment 1, rats consumed an alcohol-containing liquid diet. Blood was collected for ethanol, glucose and lactate analyses and livers were excised for lipid determination. Blood ethanol and liver fat were lower when rats consumed the high carbohydrate diet. Glucose concentrations were lower in rats fed the high fat diet compared with those fed the high carbohydrate diet when ethanol was consumed. In Experiment 2, rats consumed a high fat, ethanol-containing diet for 13 d. Half of the rats were switched to a high carbohydrate, ethanol-containing diet for an additional 11 d. The same analyses were carried out as for Experiment 1. Switching the high fat-fed rats to the high carbohydrate diet reversed the high blood ethanol and high liver fat values, even though the rats consumed significantly more alcohol with the high carbohydrate diet. In Experiment 3 the same high fat and high carbohydrate diets without ethanol were consumed for 2 wk, at which time ethanol was administered acutely, intraperitoneally, at 2 g/kg. Blood was analyzed for ethanol, glucose and lactate 30, 60 and 120 min after injection. Rats fed the high carbohydrate diet had lower blood ethanol but higher lactate at 120 min compared with those fed the high fat diet. The results suggest that the rate of ethanol elimination is slower in rats fed high fat than in those fed high carbohydrate diets, resulting in elevated blood ethanol and liver fat levels for the former.

KEY WORDS: _ blood alcohol _ blood glucose _ blood lactate _ liver fat _ rats

Figure 9.1 A facsimile of the first page of a journal article showing key initial features (*Journal of Nutrition* 132, no. 9, 2002, 2732)

with High-Carbohydrate versus High-Fat Diets." The colon allows the title to be read in segments, which makes it easier to grasp, and the new wording avoids using both "in conjunction with" and "compared with," reducing the potential ambiguity regarding the two types of diets being contrasted. Besides enhancing the readability of lengthier titles with the use of subtitles, it may be possible to eliminate some words altogether. For instance, "A Study of the Effects of" can be stated more briefly and directly as "Effects of." Or, titles that read as full sentences can be shortened by wording them headline style. In the example above, the verb "are" can be eliminated without altering the title's intended meaning.

As to the use of shortened language forms in titles, for indexing purposes it is preferable to avoid using statistical formulas, abbreviations (e.g., "AODR" versus "alcohol and other drug related"), special characters ("α-tocopherol" versus "alpha-tocopherol"), or symbols ("NaCl" versus "sodium chloride"), although some of the more common symbols, acronyms, or abbreviations (such as DNA) may be less problematic. It is also more helpful to use words rather than expressions with superscripts or subscripts ("four-carbon alcohols" versus "C_4 alcohols").

AUTHOR BYLINE AND AFFILIATION

The part of the manuscript that lists the authors and their affiliation, seemingly the most straightforward item, nonetheless requires attention to several considerations. Even the most basic question may raise concerns: Who should be listed as an author? Even when the manuscript itself is prepared only by one writer, the basic criterion for deciding whether any others should be listed as authors is the degree of importance of their contribution in planning and conducting the research and in leading to the generation of the new knowledge being reported. Though it is helpful if authorship is determined at the beginning of a project, such matters still are not always settled easily. Should a graduate student's supervisor, a head of a laboratory, or a laboratory's technician automatically be listed? Again, how important and substantial is any particular individual's contribution? How responsible is the individual for the work's achievement and how accountable for its results? By such measures, one also may need to list contributors who are since deceased, with an explanatory footnote. Individuals whose contributions do not meet a level appropriate for outright authorship may be acknowledged separately.

With multiple authors, parallel questions regarding the order of listing their names must also be addressed. While degree of contribution is a major criterion for determining author order, conventions vary by discipline or research groups. Should the head of a laboratory, the project leader, the academic adviser, or the researcher with the greatest name recognition be listed first, or listed last? How should postdoctoral fellows or graduate students be listed? As some journals and research groups do to avoid such sensitive and often complex decisions, should the authors simply be listed alphabetically? The substantial degree of collaboration in our day among researchers at different stages of their careers requires frank and open discussion, as early as possible in the research process, regarding how author credit and order are to be assigned for work that is likely to be submitted for publication.

As to the format for listing authors and their affiliations, there is some variation among publisher guidelines. The *Journal of Nutrition* requests the following:

Ex. 9.1

The names of all authors (first name, middle initial, last name) including their departmental and institutional addresses. Indicate which authors are associated with which institutions by footnotes. Identify a corresponding author and provide a complete mailing address, telephone number, fax number, and email address.

In instructions to authors given by journals at large, first or middle names of authors may or may not be spelled out, and their degrees may or may not be included. Author names may be accompanied by superscripted symbols for affiliations that are typically placed either directly below the author listing, as in Figure 9.1, or in a footnote. Note also in Figure 9.1 that the affiliation *symbols* are distinguished from the superscripted *number* used in the article's title for acknowledgement purposes. However, a numerical superscript accompanies the primary author's name (Hans Fisher) to identify him in a footnote (not shown) as the one "to whom correspondence should be addressed," noting his e-mail address but atypically omitting a postal address. Reader correspondence will include reprint requests and may be worded as such ("Requests for reprints should be addressed to . . ."). Author affiliations and the corresponding author sometimes are identified at the end of the paper, following the reference list.

PREPARING THE ABSTRACT

Examples of different types of abstracts are given in Chapter 3, including the differences between informative and descriptive abstracts. Here the discussion is somewhat extended for preparing article abstracts in particular, emphasizing the importance of consulting the guidelines for authors that journals provide. While journal guidelines have standard expectations about conciseness and following an IMRAD model in summarizing the paper, they may differ significantly in such details as length and division style (headings, paragraphing). Some guidelines also may contain greater specificity than others. The *Journal of Nutrition* gives these instructions (including keyword usage):

Ex. 9.2

The abstract must be a single paragraph of no more than 250 words summarizing the relevant problem addressed by the study and the theory or hypoth-

---

esis that guided the research. The abstract should include the study design/ methodology and clear statements of the results, conclusions and importance of the findings. Three to five key words for indexing purposes must be listed at the end of the abstract.

Some guidelines call for demarcating a paper's IMRAD components with specific headings, as shown here for the *Journal of Studies on Alcohol:*

Ex. 9.3
Abstracts should be 250 words or less and must include the following information under the these four headings: (1) *Objective:* the background and purpose of the study; (2) *Method:* the study design, setting, participants (including manner of sample selection, number and gender of participants) and interventions; (3) *Results:* details of major findings; and (4) *Conclusions:* main inferences drawn from results and potential application of findings.[5]

The *Journal of Molecular and Cellular Biology* instructs authors to refrain from using abbreviations, references, and diagrams (boldface in original):

Ex. 9.4
Limit the abstract to **200 words or fewer** and concisely summarize the basic content of the paper without presenting extensive experimental details. Avoid abbreviations and references, and do not include diagrams. When it is essential to include a reference, use the same format as shown for the References section but omit the article title. Because the abstract will be published separately by abstracting services, it must be complete and understandable without reference to the text.[6]

Abstracts should be as clear, brief, concise, and simple as possible, while staying anchored to the paper's actual content and standing independently for separate publication. It is best to write the abstract after the paper itself is finished, so that its content is more efficiently selected from a concrete and global view of the paper's various sections. In this way, it also will be easier to select the most appropriate keywords that typically must accompany the abstract, such as those that follow the abstract in Figure 9.1. The abstract's clarity and precision will also be maximized by careful use of verb tenses. Just as in the paper itself, the work is most rigorously and precisely reported using the past tense: what the authors *studied,* how they *designed* the experimental work, and what they *measured* and *observed.* An exception is the use of present tense in an abstract's introductory sentence to announce the paper's subject or purpose (such as "this paper *describes*"). Finally, the scientific clarity of an abstract's wording also will be aided when it is written with consideration of international audiences, avoiding colloquial language and being mindful of globally shared terminology.

ACKNOWLEDGMENTS

Authors of a paper customarily acknowledge those who have contributed to the work, including individuals, professional organizations, companies, and funding agencies. Such contributions include technical assistance such as data collection, advice and ideas during discussions, laboratory facilities or equipment, chemical or animal supplies, grant funding, and publishing permissions. The location in the paper for such acknowledgments varies. In the example in Figure 9.1 from the *Journal of Nutrition,* the paper's title has a superscripted number directing readers to a footnote that acknowledges commercial funding: "Supported in part by an award from Johnson & Johnson." Or, in this final endnote in the References and Notes list of a paper from the journal *Science,* one of the three authors (Udayan Mohanty) writes: "U.M. thanks S. Rice, M. Fixman, and R. Marcus for their encouragement over the years during which this research was carried out."[7] Some journals prefer that acknowledgments be placed in a separate section headed as such that follows a paper's conclusion and precedes its reference list. The *Journal of Nutrition* tells authors only: "Technical assistance and advice may be acknowledged in a section at the end of the text." The *Journal of Virology* provides authors with these more elaborate instructions:

Ex. 9.5

Acknowledgments. The source of any financial support received for the work being published must be indicated in the Acknowledgments section. (It will be assumed that the absence of such an acknowledgment is a statement by the authors that no support was received.) The usual format is as follows: "This work was supported by Public Health Service grant CA-01234 from the National Cancer Institute."

Recognition of personal assistance should be given as a separate paragraph, as should any statements disclaiming endorsement or approval of the views reflected in the paper or of a product mentioned therein.[8]

The order of acknowledgment may be either hierarchical or functional. In a hierarchical order, research supervisors and professional peers are acknowledged first, followed by technical assistants and any others who contributed. A functional order acknowledges the most valued contributors first. This example from a paper in the *Journal of the American Osteopathic Association* combines institutional and individual acknowledgments functionally, but with an atypical order by degree designations·

Ex. 9.6

We wish to thank the American Osteopathic Association and the West Virginia School of Osteopathic Medicine for their generous support of our research efforts. We also wish to express our thanks to the following students whose help has been invaluable during this study: Melissa Painter, DO, Jeff McVey, DO, David L. Prisk, MSIV, Kathleen M. Waldron, MSIV, and Amy Wells, MSIII.[9]

The following rather lengthy example—neither hierarchical nor functional strictly—acknowledges a wide range of contribution types, including field assistance, computational resources, permission to collect animals, financial support, institutional approval, and manuscript revision suggestions from colleagues·

Ex. 9.7

I thank M. Angilletta, Sr., R. Estes, and D. Kling for field assistance. A. Dunham and J. Congdon provided logistical support, advice, and encouragement. Discussions with P. Niewiarowski facilitated design and execution of the study. W. Porter kindly assisted in the use of his computer programs to calculate activity times. Lizards in New Jersey were collected with permission from the New Jersey Department of Environmental Protection, Division of Fish, Game, and Wildlife. All work was conducted with the approval of the University of Pennsylvania Institutional Animal Care and Use Committee. Financial support was received from the University of Pennsylvania Research Foundation and a Sigma Xi grant-in-aid of research. Previous versions of the manuscript were improved by J. Congdon, J. McNair, L. Rome, and R. Winters.[10]

Acknowledgment of persons requires special care to avoid any potentially demeaning ways of distinguishing individual contributions, such as by professional titles or by inconsistent uses of other references. Naturally, utmost care must be taken to spell all names correctly, whether of persons or groups.

MAIN TEXT: IMRAD STRUCTURE, STYLE, AND CONTENT EDITING

Journals typically expect papers that report original experimental work to be organized by the IMRAD conventions, and their author guidelines usually state this requirement. These prescriptions include how a submitted manuscript should be divided into sections and subsections, along with their specific wording and style. Authors also may be instructed regarding other conventions, such as those in the following guidelines from *Physiological and Biochemical Zoology* for organizing the text.

Ex. 9.8

The main body of the text should be divided into sections headed Introduction, Material and Methods, Results, and Discussion, followed by Acknowledgements, Literature Cited, Tables, and Figure Legends. These headings should be set with no indentation from the left margin. Primary subheadings should be underlined and also set with no indentation from the left margin. The first paragraph under each of these headings should not be indented; thereafter, each new paragraph should be indented. Secondary subheadings should be underlined and followed by a period, with no indentation from the left margin. The text should begin on the same line.

If the manuscript reports on work conducted on vertebrate mammals, the appropriate institutional approval number should be listed in the Materials and Methods section of the text.

Footnotes should be incorporated into the text.

Spelling may follow either American or British convention, which must be consistent throughout. Punctuation, however, should follow that recommended in *The Chicago Manual of Style*.[11]

Depending on the particular research area, as well as whether the manuscript will be submitted on paper or electronically, guidelines also may contain instructions for other textual items. These features could include preferences regarding software, typeface, fonts, letter size, spacing, pagination, built-in commands for notations (such as footnotes or endnotes and super- or subscripting), hard returns, tabs, and special characters or symbols. Additional instructions may be given for presenting statistical methods and equations; supplementing or appending data; using nomenclature, units of measure,

251

and abbreviations; and formatting of visuals (e.g., placing, labeling, sizing, coloring).

While the discussion in Chapter 8 on the IMRAD components of experimental dissertations applies just as well to scientific papers, it will be useful here to emphasize practices for content editing and conciseness that speak to their differences. In this regard, three factors become especially important. First, the degree of detail that may be expected in a dissertation sometimes is unnecessary for journal audiences. Second, given the reality that journals operate under financial constraints, features like the length of a paper, number of visuals, and use of color will affect incurred costs. Journals may pass such costs on to authors, so submission guidelines should be read carefully for these practices. Third, submission of an article to a journal involves the peer review process. Peer reviewers may request that an accepted paper be revised in ways that require authors to reword, reorganize, expand, or shorten content. Mindful of these factors, let us consider how writers of journal articles can meet expectations effectively for each IMRAD component. The discussion here will refer to a journal article by Michael Angilletta based on a chapter of his 1998 dissertation (discussed in Chapter 8) for a PhD in biology at the University of Pennsylvania. That dissertation's fourth chapter, titled "Variation in Metabolic Rate between Populations of the Fence Lizard *Sceloporus undulatus,*" was published in 2001 in *Physiological and Biochemical Zoology* under a slightly reworded title, "Variation in Metabolic Rate between Populations of a Geographically Widespread Lizard" (hereafter referred to as "Metabolic Rate"). As already noted, a widely encouraged practice with scientific dissertations is that their chapters be written for prospective publication as individual articles.

Introduction

An article's introduction, like that of a dissertation's experimental chapter, should delineate clearly and directly the framework for the reported research. This background statement should provide with efficiency and clarity the following elements:

- the nature, purview, scope, and significance of the subject;
- bibliographic references to the directly relevant work of others;
- the specific purpose of the undertaken study;
- the rationale for using the particular experimental methodology, design, and species;
- the primary results and the conclusions they suggest.

Given a dissertation's purpose of showcasing fundamental and extensive comprehension of a subject area, background statements in experimental chapters may lean toward providing more detail and bibliographic references than would be necessary for articles. An article's introduction can be shortened by an added sense of restraint to avoid references to previous work not directly related to the current research being described. When the relevant literature is extensive, citations can be strategically limited to those sources that are representative, especially important (early or seminal papers), or comprehensive (key reviews). Beyond merely acknowledging prior work, citations have the effect of tactically situating the article's author(s) within the larger community of researchers (often anticipating peer review). Both intuitively and pragmatically, citations allow an understanding of the intertextuality of scientific writings.

Article introductions typically devote the most attention and space to the first two bulleted items above—that is, the research context—and conclude with a few words that address the other aspects, particularly purpose and methodology. The introduction to Angilletta's "Metabolic Rate" article first establishes the problem (causes of geographic variation in the life history trait of maintenance metabolism) and then explains the suitability of the fence lizard for studying it, before these concluding sentences that address his methods and aims:

Ex. 9.9
In this study, I quantified the effects of temperature, time of day, and season on resting metabolic rates (RMR) of NJ and SC lizards. These data were used to estimate daily and annual maintenance metabolism of lizards in nature and to assess whether energy expenditure, as well as energy assimilation, contributes to geographic variation in the life history of *S. undulatus*.

Although most of the article introduction's content is worded as in the dissertation chapter from which it originated, some sentences were shuffled for improved logical flow, and some were added to underscore the problem's importance. In particular, to "sell" the paper to readers (and editors) more convincingly, Angilletta added the following sentences, leading into those shown in Ex. 9.9, that emphasize some key notions: "Intraspecific variation in behavior or physiology must contribute to the difference in production between NJ and SC lizards. Indeed, lizards from South Carolina have a greater rate of metabolizable energy intake than lizards from New Jersey at the preferred body temperature (Angilletta 2002). It is not known whether annual maintenance metabolism differs between NJ and SC lizards."[12]

The vast majority of introductions neglect any mention of results and conclusions, making readers wait to get a full sense of the work's outcome and significance. This is akin to the journalistic or literary tactic of keeping readers in suspense until the end, but is best avoided in scientific articles. The following final paragraph from the introduction of another study—on ameliorating alcoholic liver pathology by inhibiting cytochrome CYP-2E1 activity—does include a sense of what the results suggest (emphasis added).

Ex. 9.10

To investigate the effect of ethanol-induced CYP2E1 on liver fatty acid composition and liver pathology, we analyzed livers from rats treated with ethanol in combination with inhibitors of CYP2E1 induction. Some results reached in previous studies of these livers (Morimoto et al. 1993, Morimoto et al. 1994, Morimoto et al. 1995, Rouach et al. 1994) are given for the purpose of correlation. *The data indicate an important role for CYP2E1 in the ethanol-induced changes in hepatic fatty acid composition.*[13]

Given that introductions do refer to the work of others as well as to the new research being reported, one other aspect of introductions to check carefully is the use of tense, to avoid ambiguity or inconsistency. An article's introduction (or other parts) may use various tenses, depending on what is being mentioned. When introducing the paper itself, it is appropriate to use the present tense ("This report *describes*" or "This study's purpose *is* to"). However, when the author's research is being introduced, past tense is needed (as in Ex. 9.10: "To investigate . . . we *analyzed* livers"). Similarly, present rather than future tense is more suitable when mentioning the paper's content ("A new technique *is* described," not "*will be* described"). Tense also may vary in references to the literature. Present tense may be used for established findings ("C57 mice *select* ethanol over water when . . ."), but reference to specific findings calls for either the past or present perfect tense ("McLearn *found*" or "Ethanol *has been shown* to"). Tense distinguishes a universal truth from a finding in the particular case.

Materials and Methods

The materials and methods section of a paper is a straightforward factual description of what the investigator(s) did to obtain the results that are presented and discussed in succeeding sections. Without a complete, precise, and accurate picture of a study's methodological details—experimental design, instruments, techniques, procedures, measurement—readers cannot properly or fully evaluate the outcome and its significance. Instructions to authors in some journals include specific requirements for particular aspects of methodology. The *Journal of Nutrition,* for instance, spells out expectations regarding how authors should provide their protocol in human and animal research (e.g., ethics committee approval, subjects' characteristics), explain their use of statistical methods (specific tests, data representation), and describe formulation of diets. The journal's instructions include references to manuals and articles that provide further details on standardized expectations, such as the following guidelines for "Diets."

Ex. 9.11

Composition of control and experimental diets must be presented. When a diet composition is published for the first time in *The Journal of Nutrition,* utilize a table or a footnote to provide complete information on all components. If previously described in *The Journal,* a literature citation may be used. The proximate composition of closed formula diets should be given as amounts of protein, energy, fat, and fiber. Components should be expressed as g/kg diet. Vitamin and mineral mixture compositions should be included using *Journal of Nutrition* units and nomenclature. For a discussion of the formulation of purified animal diets, refer to Baker *(4)* and to a series of ASNS publications *(5–8).*

Some manuscript preparation guidelines, including those in the *Journal of Nutrition,* have additional sections stating their expectations for using standard units of measure, nomenclature, and abbreviations. As emphasized in Chapter 8, on scientific dissertations, full details of the overall experimental methods must be provided so that other researchers can repeat the work and obtain parallel results. Generally, a citation will suffice when the details of methodology already have been published elsewhere, unless there are modifications to report. For instance, the "Data Analysis" subsection of Materials and Methods in Angilletta's "Metabolic Rate" article contains various references to established analytic procedures, including this simple one: "CO_2 was used to estimate energy expenditure, using the appropriate conversion factor for the respiratory exchange that was observed (Nagy 1983)." Rather than exhaustive detail or familiar minutiae, the *principles* on which the methods are based often are more important to explain, as Angilletta notes here:

Ex. 9.12

I assumed that lizards would not become active each day until surface soil temperature reached a minimum of 20°C. For each month, I totaled the hours that *S. undulatus* could achieve its preferred body temperature (33°C; Angilletta 2002) per day after the onset of activity. By summing these hours over an entire year, I arrive at an estimate of annual activity time.

The author simply explains how he used the concept of "preferred body temperature" to calculate the "annual activity time" of his lizards.

Writing and organizing a study's materials and methods section should, as Katz has advised, "be as close to an algorithm (a computer program) as possible; so work in an absolutely lean and spare style, and break the overall section into many clearly labeled parts that fit into a straightforward outline."[14] The materials and methods section of Angilletta's article is not only streamlined in its detail but also parceled into subsections that describe "Animal Collection and Care," "Measurement of Metabolic Rate," "Data Analysis," and "Activity Time and Maintenance Metabolism." The dissertation chapter version of the methods write-up contains an additional segmentation into "Regression Models," not deemed necessary for the article, its single paragraph now merged with the data analysis subsection.

Making changes in the methodology section of a paper—such as in sectioning, content, or wording—may be something that peer reviewers insist on the author doing. Revisions in wording may involve, for instance, making the language more direct, precise, or comprehensive in reflecting what was done. Consider the following two sentences by Angilletta, the first version from his dissertation and then the revised version, with the three changes italicized, from his "Metabolic Rate" article.

Ex. 9.13
1. Because the experimental design involved repeated measures of the metabolic rates of individuals, analysis of covariance (ANCOVA) with repeated measures was used to examine the between-subjects effect of population on metabolic rate.
2. Because the experimental design involved repeated measures of the

---

metabolic rates of individuals, *ANCOVA* with repeated measures was used to examine the *among*-subjects effects of population *and season* on metabolic rate.

Each of these simple changes had its own rationale—one based on readers' familiarity with the term ANCOVA (obviating its being spelled out), another for grammatical precision, and the third to more accurately reflect the key parameters. An example of a substantial change in methodological content requested by peer reviewers in Angilletta's article manuscript is the addition of a section (not present in his dissertation chapter) titled "Activity Time and Maintenance Metabolism," which describes "a biophysical model to estimate the maximum duration of daily and annual activity of lizards." To meet the reviewers' new calculation requests with his added measurement model, Angilletta used special software with the assistance of a colleague who is recognized in the article's acknowledgements section ("W. Porter" in Ex. 9.7). Such changes in methodology naturally also require accompanying adjustments in how data are presented in the results section.

Results

The heart of an experimental article is its presentation of results, or the new information to be gleaned by readers. However, while the results section is the most important part of an article, if the section on methods and materials has done its job thoroughly it likely will also be the shortest part. One feature of the results section that permits greater verbal economy is the use of visuals—whether photos, graphs, or tables—to which the author can refer readers for the complete picture. As in the article's introductory and methodological content, authors must apply strategies here not only for clear, logical, and smooth reporting of findings but also for selectivity of detail. Some journals provide more detailed instructions than others for presenting results, as do these explicit guidelines from the journal *Biochemistry:*

Ex. 9.14
Results. The results should be presented concisely. Tables and figures should be used only if they are essential for the comprehension of the data. The same data should not be presented in more than one figure or in both a figure and a table. As a rule, interpretation of the results should be reserved

for the Discussion. In the interest of economy of space, Supporting Information (also subject to review) should be submitted as a separate file. The policy of the Journal is to publish only representative data. For example, routine gels and linear plots will not be published. However, if such information is important for evaluating the research, it should be included in the submission for the reviewers or as Supporting Information. The Supporting Information will be included in the World Wide Web edition of the Journal (see Supporting Information).[15]

So long as readers are given a clear understanding of the criteria used in the selection process, it will suffice, as *Biochemistry* instructs, to report the most representative results. In reporting data for an alcohol drinking study with mice, for instance, one may indicate that extremely irregular consumption data were discounted due to apparent spillage, rather than ingestion, caused by how animals made contact with the fluid containers. This statement from a study on alcohol's effects on rat muscle proteins explains why certain data were omitted.

Ex. 9.15

Body weights are not presented because it has been previously shown that in glucose-fed control and alcohol-fed rats, body weights can change markedly over a day because of episodic engorgement of the liquid diets, so that body weights are not meaningful (19). Alcohol has been previously shown to reduce muscle weight (6, 20).[16]

In some cases, one may also report the smaller percentages of atypical observations, explaining why they are not representative, and then focus on the "best" results.

The findings should be reported concisely and directly, each paragraph beginning with a topic sentence that indicates the larger picture and leads to progressively more specific detail. The following paragraph from Angilletta's "Metabolic Rate" begins with sentences that announce the overall nature of the data reported on his lizards' metabolic "budgets" (in terms of RMR, or resting metabolic rate), proceeds to specific findings, and concludes with broader statements again on the lizards' energy "expenditures."

Ex. 9.16

I compared daily and annual maintenance budgets of NJ and SC lizards, calculated from regression models of RMR. On average, temperature and body mass explained 70% of the variation in RMR (Table 3). In all seasons, temperature and body mass were better predictors of RMR during scotophase than RMR during photophase. Based on the biophysical model, annual activity times were 1,802 and 2,387 hours for NJ and SC lizards, respectively. Assuming that activity of lizards in New Jersey and South Carolina corresponded to the predictions of the model, SC adults had a maintenance expenditure of 45.8 kJ, whereas NJ adults had a maintenance expenditure of 53.7 kJ. Note that the annual activity of SC lizards was estimated to be 32% greater than that of NJ lizards, but annual maintenance expenditure was estimated to be 15% less. The relatively high maintenance expenditure of NJ adults was caused by the significantly higher RMR of NJ adults during summer and fall. Adult lizards in New Jersey have greater maintenance expenditures than those in South Carolina (Table 4), and this results in a greater annual maintenance budget.

The paragraph's content is shaped like an hourglass, or egg timer, with the broader beginning and ending sentences funneling toward the specific data in the middle. While the paragraph's topic sentence alludes to both daily and annual metabolic data (RMR), only the annual data are reported there, with the daily data reported in the following paragraph. The results that matter most are those that provide a direct answer to the research question. One must avoid the temptation to present every single finding, exhaustively and indiscriminately, which also can lead to an excess of accompanying visuals. Laboratory notes, and sometimes even a dissertation, may contain more methodological details, collected data, and graphics than readers of an article will need to know. Therefore, there must always be some degree of personal selectivity in narrating what was observed in the laboratory and in deciding which parts of the findings to present.

Authors should consult a journal's manuscript guidelines regarding the specific manner of presenting data or incorporating visuals. This may involve how authors use symbols, units of measure, or nomenclature; display notations and equations; design, label, and cite tables or figures; and explain computations. In making computations, for instance, authors typically are instructed to include sufficient definitions of any particular models used in deriving data, including references to the appropriate sources in the literature, for validation by readers. In some cases, authors may wish to take advantage of supplemental forms for presenting results. The following statement in the *Journal of Chemical Physics* informs authors of an online site where they can place certain types of article materials.

Ex. 9.17

Electronic Physics Auxiliary Publication Service (EPAPS) is a low-cost depository for material that is supplemental to a journal article. Appropriate items for deposit include multimedia (e.g., movie files, audio files, 3D rendering files), data tables, and text (e.g., appendices) that are too lengthy or of too limited interest for inclusion in the printed journal. Retrieval instructions are footnoted in the related published paper. Prominent links in the online journal article allow users to navigate directly to the associated EPAPS deposit. EPAPS deposits may also be retrieved by users free of charge via command-line FTP or via the EPAPS homepage. Authors are encouraged to deposit multimedia files with EPAPS.[17]

For visuals that are to be printed with the paper, authors also must follow journal-specific instructions for how to submit electronic copies along with camera-ready artwork. Certain file types may be required and others prohibited, for instance, as in these instructions from *Physiological and Biochemical Zoology:*

Ex. 9.18

Figures for the accepted manuscript should be sent in the following formats, on the same floppy disk as the final manuscript. Camera-ready hard copies MUST be provided along with the electronic file.

    Line art should be provided as bitmapped .tif files saved at a resolution of 800–1200 dpi (pixels per inch)

    Black and white photographs, micrographs, etc. should be provided as grayscale .tif files saved at a resolution of approximately 300 dpi.

    Color art should be provided as .eps files, CMYK, at a resolution of 150–300 dpi. (If this format is not available, provide color art as Photoshop files.)

    The following formats are not acceptable for figures: Word or Powerpoint files, .jpeg, or .gif files.

Separate sections in the instructions of that journal provide additional details for preparing tables as well as camera-ready artwork, the latter including details on quality, font, content, photographs, Internet graphics, and copyright.

The results section of a paper can be kept as lean as possible, then, not only by reporting just the data necessary for answering the research question, but also when appropriate by the use of electronic sites for posting supplemental material. Another way to be economical is to avoid repeating in the text information that is provided in tables or in legends to illustrations. Beyond contextual or pinpointed commentary on the data, one can simply refer parenthetically to the sequentially numbered visuals. At the same time, authors should minimize repetition in a figure's legend of information, such as methodological details, already given in the text. The key is a space-saving balance. Recall also that manuscript reviews calling for changes in methodology, such as those involving measurement or data analysis, may require parallel revisions in how the data are reported. In Angilletta's "Metabolic Rate" manuscript, when peer reviewers requested that the author calculate the lizards' *daily* and *annual* patterns of energy utilization, not just the *seasonal* patterns reported originally, the paper now had to report not only how he did those new calculations but also the newly derived data. This required the addition of two paragraphs (one of which is shown in Ex. 9.16). Peer reviewers also asked for changes in Angilletta's visuals. Their request for new data analyses required adding a bar graph, modifying an existing table, and (due to data now deemed unnecessary), removing a table used in the dissertation write-up of the study.

Discussion

After presenting the results, authors must analyze them in a closing discussion—within the study's hypothetical framework—to arrive at the conclusions and implications permitted by those findings. Since there is no standard way to write the discussion section, giving authors a freer hand than in the preceding sections, it is the most challenging part of the paper to write. However, it is logical to begin with a brief reminder of the most important findings and then to move outward toward progressively more generalized statements about the relationships among the data, the connections between the present findings and those of other researchers, and ultimately the principles and inferences that are applicable universally. Precisely because authors do have such a free hand here, they must be particularly disciplined to avoid verbosity and wandering that will work against their keeping anchored to a small number of key points worthy of development and special emphasis. The discussion is a chief target among reviewers for recommended trimming, and it can determine whether a paper is accepted or rejected. An effective discussion must be grounded in the actual findings and when appropriate should refer to the related work of others.

The following are fundamental questions that a discussion should address, along with excerpted examples in brackets from Angilletta's "Metabolic Rate":

- *What major patterns do the data exhibit?*
  ["*Sceloporus undulatus* exhibited a marked diel cycle of RMR, similar to those of other diurnal lizards. Metabolic rate was maximal at 1200–1930 hours and decreased significantly at the onset of scotophase."]
- *How do the methods, results, and interpretations compare with those of others?*
  ["Metabolic rates for *Sceloporus undulatus* reported here compare favorably with those reported by other investigators. For example, I observed a RMR of 5.4 ± 0.9 J $g^{-1} h^{-1}$ for lizards at 33°C. Zannoni (1997) reported a RMR of 5.8 ± 0.7 J $g^{-1} h^{-1}$ . . . It is important to note that that RMR reported here and by Zannoni are not equivalent to SMR because measurements were made during periods of photophase as well as scotophase. Also, neither study selectively reported minimum rates of metabolism observed, as in many studies of SMR (e.g., Feder and Feder 1981; Tsuji 1988; Rowe et al. 1998)."]
- *Are any discrepancies with others' results or unresolved issues addressed?*
  ["The discrepancy in patterns of acclimatization observed in sceloporine lizards is not surprising because seasonal acclimatization is a complex phenomenon driven by a multitude of metabolic processes (Clarke 1993). . . . Clearly, simple hypotheses cannot adequately explain patterns of acclimatization in ectotherms, and studies should be designed to examine hypotheses that incorporate multiple causality at a biochemical level (Clarke 1993)."]
- *Do the data allow important extrapolations or predictions?*
  ["Most important, my regression models seem capable of predicting maintenance expenditure with considerable accuracy. This can be demonstrated by comparing daily maintenance expenditure predicted by regression equations to that determined by the doubly labeled water method. Joos and John-Alder (1990) used doubly labeled water to . . ."]
- *How do the findings provide answers to the posed problem?* (Ex. 9.9)
  ["Very little is known about the contribution of variation in energy expenditure to geographic patterns of life history in *S. undulatus* . . . The lower

maintenance cost of SC lizards necessarily results in a higher production efficiency. Thus, the greater output of SC lizards is partly a product of physiological differentiation between populations."]

• *What conclusions, generalizations, and implications do the results allow?*
["My results highlight the importance of considering multiple causality of ecological phenomena (Quinn and Dunham 1983). Intraspecific variation in life history phenotypes of *S. undulatus* results from differences in energy assimilation and energy expenditure between populations. Similar physiological mechanisms are likely to operate in other species that are geographically widespread. Undoubtedly, a genuine understanding of the proximate causes of geographic variation in the life histories of ectotherms will not be achieved by formulating simple causal hypotheses but will compel an integrative approach designed to tease apart the relative contribution of multiple mechanisms."]

The conclusions articulated in a discussion must be supported closely by the specific observations and the relevant patterns in the data. For instance, Angilletta cogently supports one of his conclusions—that "the primary cause of the diel cycle of RMR observed in this study was probably an environmental cue, rather than a circadian rhythm"—by concisely enumerating his relevant methodology and data patterns:

Ex. 9.19

First, lizards were allowed to habituate to the chamber for 2h before measurements. Second, RMR did not decrease significantly until the onset of scotophase each day. Finally, the difference in RMR between photophase and scotophase appears to be as pronounced on the fourth day of measurements (36°C) as it was on the first day (33°C).

A paper naturally must end with an overall conclusion, though it need not necessarily be labeled or subheaded as such. Once all the important results have been analyzed, the author should stand back and take a moment to reflect upon what are the most important synthesizing thoughts about the study that could be left to readers. As an example, the concluding paragraph of Angilletta's article appears with the final bulleted question for a discussion section, above ("My results highlight the importance of . . .").

A final point that bears emphasis is that Angilletta's overall discussion demonstrates various ways to be concise. He focuses on those few points that are most significant, without excessive explanation, as in Ex. 9.19 and in the preceding excerpts that address basic discussion questions. He also is concise stylistically, with liberal use of the active voice ("I observed") as well as by avoiding wordy constructions (such as "it is interesting to note that"). As a space-saving revision recommended by reviewers, he also removed subheadings typically used for the different kinds of results being discussed, originally included in his dissertation version ("Thermal Sensitivity," "Diel Variation," "Seasonal Acclimatization," "Variation between Populations"). Simplicity and economy in expressing the meaning of one's scientific labors commands more authorial power and reader attention than verbosity and eloquence.

ETHICAL PUBLICATION IN SCIENCE

No discussion of the preparation and publication of a scientific article can be complete without mention of the ethical responsibilities of those who are involved in the process. Ethical behavior in the publication of scientific research receives more scrutiny today than ever before. Authors of papers, reviewers, and publishers must abide by certain established professional codes, some of which are in writing (copyright laws, for example), and others that are understood. Violations of these expectations, witting or not, can carry a heavy personal and professional price. Scientific journals and associations have disseminated detailed statements on publication ethics. *On Being a Scientist: Responsible Conduct in Research,* a booklet published jointly by the National Academy of Sciences, the National Academy of Engineering, and the Institute of Medicine in 1995, includes a section on authorship practices that covers ethical responsibilities. In addition, the Office of Science and Technology Policy (OSTP) in the US Department of Health and Human Services has proposed a common definition of scientific misconduct to be followed by all federal agencies. That proposal, announced on October 14, 1999, defines research misconduct as "fabrication, falsification, or plagiarism in proposing, performing, or reviewing research, or in reporting research." Finally, the American Chemical Society's *Style Guide* for authors and editors includes an appendix, "Ethical Guidelines to Publication of Chemical Research," which is also incorporated in some form in the author guidelines of many scientific journals, such as those affiliated with the American Institute of Physics and the American Geophysical Union. Formal attention to ethics in scientific publication is also given during graduate study, though arguably not enough.[18] There are standard sets of specific ethical rules that apply to authors, reviewers, and journals. The ethical rules listed in the following sections are representative of widely held expectations, and are given more detailed articulation in *The ACS Style Guide.*

OBLIGATIONS OF AUTHORS

Ethical conduct on the writer's part relates to such matters as authorship credit, originality of research, multiple publication, treatment of sources, and intentionally misrepresented data. The latter is so infamous a fraud that, when discovered, it is widely publicized. Luria emphasized that "except for a few emotionally disturbed individuals, people in science do not cheat"; although he also noted that "science does not select or mold specially honest people: it simply places them in a situation where cheating does not pay."[19] Authors of scientific works are generally expected to adhere to the following list of principles. These expectations may seem fairly obvious, and they are widely understood and followed by seasoned researchers, but making them explicit here will serve as a convenient benchmark for those who are embarking on a scientific life.

- Present the research accurately and fully, allowing peers to replicate the results.
- Cite relevant research judiciously, emphasizing influential publications that give readers a historical sense of the work, and avoiding unnecessary references.
- Identify the source of all quotations, and maintain the confidentiality of information obtained through personal contacts or professional roles (such as peer reviewing) without permission.
- Alert readers to any unusual risks associated with the reported research, such as with substances, instruments, procedures, living specimens, or phenomena.
- Consider journal space a commodity to be used sparingly, so not only be concise but also avoid fragmented publication, which wastes time in literature searches.
- Do not submit the same manuscript for multiple and simultaneous consideration by different journals, and inform the editor of related manuscripts submitted elsewhere.
- List as co-authors only those who have contributed to the work significantly, are accountable for its results, and have assented to co-authorship.
- Refrain from personal invective against of the work of others, versus a reasoned critique that adheres to professional etiquette.

- Reveal any potential conflict of interest, such as consultative or financial, associated with publication of the manuscript's information.
- Obtain permission from the copyright owner when republishing graphics or substantial parts of a paper, and then give credit for that material properly and fully.

The items on this list (and those that follow for reviewers and editors) are stated in their barest form, as simple points of reference, and any of them certainly can be elaborated on with additional context, situational contingencies, and examples. Collectively, these basic ethical expectations can be thought to make up a credo for scientific researchers who write for publication.

Of all the ethical violations that can be committed by scientist-authors, the most serious involve communicating misrepresented data and plagiarizing. Setting forth fraudulent data is rare, but researchers must also guard against a failure of objectivity due to personal or cultural bias. Seeing experimental results through the prism of expectation or of cultural beliefs can cloud the observation of what actually is the case, unduly limiting the scope of interpretation. Our time is not immune to the kinds of prejudices, for instance, that Gould exposed in the comparative craniometry of Broca and his followers, who fabricated data to reflect the culturally entrenched view that women's intellect was biologically inferior to men's. Whether the publication of scientific work in question involves a failure of objectivity or a deliberate fabrication of results, its ripples across the scientific community will waste precious time and effort to set straight.[20]

The other grave act of dishonesty in scientific (or in any) publication, plagiarism, takes various forms. One type of transgression is to copy another person's writing, word for word and without quotation marks, and represent as one's own. It is also improper to mix pieces of another writer's wording with partial paraphrasing, the latter often being minimal. A third form of plagiarism is paraphrasing or using another's *ideas* without giving credit. Ignorance regarding proper citation or use of quotation marks cannot serve as an excuse for such unethical practices.

Sensitive personal or even legal situations can arise regarding the allocation of credit for the work or ideas of others when a paper's authorship is at stake. Such cases may involve researchers of differing status, such as laboratory heads, principal investigators, postdoctoral fellows, and graduate students. Open discussions of credit allocation and authorship early in the research process are likely to prevent such conflicts. That authorship may be perceived as insufficient credit is illustrated by a prominent case: Jocelyn Bell, a physics graduate student who discovered pulsars, was listed by her thesis adviser Anthony Hewish as a co-author of the paper announcing the discovery, but the Nobel Prize was awarded to Hewish only.[21] Should Bell have been a co-recipient of that highest of scientific prizes? Within a paper itself, citing the relevant publications of one's peers is also ethically important, for so much of the reward system in science (such as grants, job status) relies on recognition of one's labors.

Clearly, ethical violations or errors in science range in type and situation. Honest mistakes from human fallibility that are reported promptly to the journal that originally published the mistake will be seen as far less troublesome than errors due to negligence. Doing research in a substandard manner, due to carelessness, impatience, or "inconvenience" of protocol, can undermine a scientist's reputation and elicit a much less merciful judgment. Finally, ethical violations that are due to outright *misconduct*—including fabricating data, misreporting results, and plagiarism—are subject to the harshest punishment, including destruction of one's career. Such egregious violations, or even just *perceived* transgressions, ultimately even risk damaging the public's perception of the scientific community as a whole, particularly when they reach the news media.[22] Research institutions that receive public funds typically have administrative mechanisms in place, including designated ombudspersons, committees, and appropriate recordkeeping, that address allegations or actual findings of ethical violations. All parts of the research enterprise, from individual researchers and academic institutions to scientific associations and the private sector, share in the responsibility to maintain a culture of awareness, prevention, and corrective action regarding potential or actual violation of ethical standards in scientific investigation and publication.

OBLIGATIONS OF MANUSCRIPT REVIEWERS

Peer reviewing of article manuscripts is a vital part of the publication process as well as of the social and collaborative nature of scientific work. Publishers must be assured of the quality and value of the submissions they receive, and the scientific process itself requires the willingness of all researchers to help scrutinize what is to be disseminated. Beyond the professional ethic of doing one's share in assessing colleagues' contributions for inclusion in the limited and costly space available for periodical issues, scientists who serve as reviewers have the ethical duty to:

- Review only those manuscripts that they feel qualified to evaluate.
- Assess a manuscript objectively, from its experimental narrative and theoretical framework to its conclusions, compositional rigor, and individuality of thought.
- Recognize and inform the editor when review of a manuscript poses a conflict of interest, such as with their own related work or personal relationships with authors.
- Maintain manuscript confidentiality, and reveal any advisory consultations.
- Explain and support bases of judgments sufficiently, and note with citation if the reported results or ideas had already been published.
- Ascertain, without being self-serving, whether the author cites the relevant work of others, and keep alert for concurrent submission of similar work to other journals.
- Complete the review promptly, or decline to do it if delays are anticipated.

In essence, reviewers should give prompt and unbiased consideration to all manuscripts, judging each contribution on its own merits and without discrimination based on race, religion, sex, nationality, seniority, or institutional affiliation. While reviewing manuscripts is a time-intensive task, the process of scientific inquiry cannot be held to its high and uncompromising standards without the willing participation of every researcher.

OBLIGATIONS OF JOURNAL EDITORS

Many of the ethical obligations of manuscript reviewers apply as well to journal editors. These shared expectations include considering all manuscripts objectively, confidentially, promptly, and with respect for the authors' intellectual independence. In addition to these overlapping ethical responsibilities, editors have either the discretionary authority or the obligation to:

- Take into account a manuscript's relation to others submitted concurrently or previously by the same author(s).
- Exercise responsibly the decision to accept or reject a manuscript, normally with reviewer advice unless the submission is inappropriate for the journal.
- Disclose or publish titles and authors of manuscripts after they are accepted, with any further disclosure of content only with permission of authors.
- Avoid conflicts of interest by delegating to other editors any manuscript they submit to their journal as editor-author, or when considering work related to their own.

• Facilitate publication of a statement that discloses and if possible corrects erroneous information in an article published in the editor's journal.
• Select manuscript reviewers carefully, choosing those who are best qualified, with the option of considering the expressed preferences of authors.

Since the decision to accept or reject a manuscript lies squarely with the editors, they have a special responsibility to make sure that the consideration process is thorough, unbiased, and advised by the most qualified reviewers possible. When editors receive a manuscript of a type or on a subject that does not fit their journal, they will do well to recommend alternative journals to which the authors may submit their paper.

With the tens of thousands of scientific journals being published, it is daunting enough for readers just to keep updated in their areas of interest. This task must not be complicated further by obfuscations or setbacks resulting from any untoward temptations associated with the relentless pressures of the "publish or perish" syndrome. Journal readers therefore are heavily dependent on the unwavering adherence to codes of ethical conduct by authors, reviewers, and publishers alike. Indeed, the very fabric of the scientific enterprise cannot remain intact without a systemic collaboration that acts to protect and nourish its unique form of inquiry and of publication.

FINAL CONSIDERATIONS ON THE SCIENTIFIC PUBLICATION PROCESS

Writing and publishing a scientific article requires not only close and rigorous attention to the content and wording of various parts of the IMRAD structure, but uncompromising adherence to a code of professional ethics among authors, reviewers, and editors. In addition to the larger picture involved in writing the article itself and in communications with others involved in the publication process, authors must be aware of the seemingly smaller but nonetheless significant details that are standard parts of the process. These details include submitting the requested number of manuscript copies, writing an accompanying cover letter, and signing publication and copyright transfer agreements. Authors should also understand whether there will be page or color charges, and even whether such charges can be minimized by posting supplemental material such as graphics on designated electronic sites. Neglect of any of these aspects can delay or even derail the publication process. Following publication itself, authors may wish to order reprints of their articles so they can extend the professional courtesy of making them available to colleagues (or even students) who request copies. Finally, the overriding concern with publishing articles in journals should not obscure the fact that the same high standards of scientific professionalism must be followed when scientists publish for lay audiences. When writing for the public, scientists must remain objective and accurate. It is also appropriate in popular writing for scientists to "translate" technical knowledge using common language rather than the codified terminology and phrasing of insiders. Even in writing for the public, however, scientists must remain within the bounds and claims of information and ideas that have been tested through dissemination in the professional literature of their field.

SCIENTIFIC GRANT PROPOSALS

Our concern here is not so much with the history as with some of the consequences of the idea itself.
—Harold J. Morowitz, *The Wine of Life*

WHAT IS A SCIENTIFIC GRANT PROPOSAL?

The sophistication and expense of experimental research requires many scientists to compete for external funding by submitting grant applications that
attempt to convince a small audience of evaluators that their proposed project
is worthy of financial support. The granting bodies may be private or public—
mainly government agencies, corporations, and foundations. Government
agencies—including the National Science Foundation, the National Institutes
of Health, the US Department of Agriculture, and NASA—collectively grant
billions of dollars annually for scientific research, but the statistical truth is
that many more proposals are submitted than can possibly be funded. The very
survival of a particular avenue of scientific inquiry may entail a rather Dar-
winian competition for a limited pool of funds. The initial submission of a
grant proposal faces tough odds for receiving support, and its authors will
likely have to revise and resubmit it in hope that it may fare better next time.
Given this intense competition, a successful proposal must build a strong case

for the merit of its research idea. To express that idea effectively, it is helpful to keep in mind that, according to one authority, "the principles underlying the writing of a journal article apply to writing a research proposal: in both, the objective is a logical, clear and succinct phraseology, and flawless reasoning."[1] The unique challenge for proposal writers is that, in contrast to reporting research already completed, they must set forth a compelling scenario and plan for research that is yet to be done.

To imagine, invent, and then communicate—indeed to argue for—a project convincingly requires much patience and thought. It is a social process of give and take among collaborating researchers, peer evaluators, and grant officials to create a document that meets high expectations in the quality of its ideas, its written expression, and its perceived promise. The inherent messiness of this process, of collaborative thinking and creativity, ultimately must yield a relatively smooth experimental narrative that argues successfully for the importance of its anticipated outcome. Applications written in haste will reflect just that to their evaluators. "Yes, true enough," concedes Montgomery, "the hurried nature of many proposals reflects the conditions and means of their production—the dash to make deadlines, to invent details, to coordinate pieces by different groups, and to sew the whole together in Dr. Frankenstein fashion so that the result walks and talks without too many seams showing and without becoming a danger to its makers. Your proposal must indeed hide all this reality." While it is most important that the project idea itself hold up to scrutiny, the proposal must also be well written *and* follow the granting body's detailed instructions regarding content, organization, and format. Sufficient time must be taken to make sure all of these elements work together to produce a thoroughly professional and compelling document. The preparation and review process for grant proposals naturally also is subject to professional standards and ethical obligations that parallel those for article manuscripts (discussed in Chapter 9).[2]

Because the overwhelming majority of scientific research grants are from federal sources, the focus here is on the conventional parts of a government grant proposal and how it can be written to maximize the chances that the envisioned research will be supported. Compared with government agencies, philanthropic foundations and corporations generally are capable of considerably less financial support, and rather than soliciting proposals they are more likely to offer resources through personal contacts. Corporate sponsors may grant such limited support as small cash awards or donation of laboratory equipment. Other private organizations with special interests, such as the National Geographic Society or the Audubon Society, may also give limited funding without solicitation of proposals. Private and commercial organizations typically provide resources under conditions that further their own narrow mission or financial interests. Government agencies, however, are expected to have the interests of the broader public in mind and can support a broader spectrum of research goals. Guidelines for preparing government grant proposals, as well as listings of the kinds of projects that interest particular agencies, generally can be found online.[3]

GUIDING PARAMETERS IN PREPARING A PROPOSAL

In creating a feasible scenario for future work, applicants must tell a likely story with an ending and consequences that can only be imagined. What good reasons are there for a sponsor to take a risk with a particular proposal? Although a grant proposal shares with other types of research proposals (such as for dissertations) the aim of selling the promise of an idea to its readers, the financial stakes make it a palpable competition. Authors of grant proposals must take stock of their personal research program to provide a clear picture of their proposed project, from its significance and approach to realities dealing with timelines and costs. A successful proposal convinces its evaluators that each of the following elements is well conceived.

- *Aims.* What is the project's goal? What are the key questions? What is the approach? Is the relevant science rigorous and of high quality?
- *Importance.* Is the project worthwhile? How innovative is the idea, approach, or methodology? Is it likely to produce results of special consequence?
- *Authority.* Is the principal investigator (and any collaborators) reliable and trained to tackle the project? Has the literature been adequately researched?
- *Feasibility.* Are the goals likely to be met if the work is conducted as described? Have the project's scope and timetable been realistically considered?
- *Expression.* Is the proposal highly readable? Is its wording as direct and clear as possible? Does it flow smoothly and logically? Are helpful visuals used?
- *Budget.* How much will the envisioned work cost? Is the requested funding truly essential to achieve the desired ends?

There is no comprehensive set of rules for writing successful grant proposals, but close attention to these kinds of questions will maximize the chances that the writing, review, and revision cycles will yield a convincing presentation that merits support. Ultimately, the quality and rigor of the proposal reflects not only the significance of the idea itself but also the professional standing that the authors personally bring to it. The overall process is complex and not always free of controversy, as with any other human endeavor involving performers and judges in a competition. As a buffer against potential conflicts of interest, granting agencies may give applicants a voice in selecting reviewers. The National Science Foundation, for instance, instructs its grant applicants that they "may include a list of suggested reviewers who they believe are especially well qualified to review the proposal [and] also may designate persons they would prefer *not* review the proposal, indicating why." Although various guidebooks on grant writing are available, and applicants can look at successful examples from their colleagues, in the end the process is a lonely and uncertain one in which proposal writers are left to their own devices as they meticulously follow a sponsor's detailed guidelines and take their chances alongside their competitors.[4]

CONVENTIONAL PARTS OF A GRANT PROPOSAL

While there is some variation in the requirements across sponsors, grant proposals typically include the following conventional elements.

- *Cover sheet:* Designates grant program and subfield; lists investigators' names, degrees, and affiliations; provides project title; specifies funding amount and duration; and identifies animal or human subjects.
- *Summary:* Describes what investigator(s) will do if the project is funded, including the problem, objectives, methods, scholarly merit, and anticipated consequences.
- *Full proposal:* Presents complete details of the project, including its relation to work in progress, extended goals, current knowledge in the field, other funded work, and how the results will be disseminated as well as their expected societal benefits.
- *Literature cited:* Lists all the sources that are cited in any part of the proposal, following a conventional and complete format.
- *Biographical sketches:* Provides professional information for all senior proj-

ect personnel (project director, faculty), including their degrees, postdoctoral training, appointments, publications, and activities—such as teaching
and service—that demonstrate their broader contributions.

- *Budget:* Specifies and justifies estimated expenses for the project, including
  salaries (senior personnel, assistants, clerical staff), equipment, supplies,
  travel, and services (e.g., dissemination, consulting).

The detailed picture of the proposed work and its scientific value must be presented clearly, convincingly, and in a highly concise manner dictated by strict
space limitations. Moreover, unlike journal articles, the proposal must be written in a manner that is accessible not only to fellow specialists but, insofar as
possible, also to technically literate lay readers—from the project's title and
summary to its full description. The conventional parts of a complete description of the project include an introduction and overview, results from prior
grant support, background information, the work's significance and applications, preliminary work done, methodological details, expected outcomes,
data analyses, special issues, and the broader consequences of the project's anticipated results. There also are *revision* considerations that apply to resubmission of a proposal that did not fare well initially.

In addition to following a sponsor's guidelines regarding content and page
limits, applicants must also conform to format prescriptions for margins, spacing, and letter size. Such requirements are intended not just to level the playing field, but also to allow for easier reading. Here, for instance, are the very
specific margin and spacing requirements for proposals submitted to the National Science Foundation.

Ex. 10.1
1. The height of the letters must not be smaller than 10 point, unless otherwise specified in the program solicitation to which the proposal is being
   submitted.
2. Type density, including characters and spaces, must be no more than 15
   characters per 2.5 cm. For proportional spacing, the average for any representative section of text must not exceed 15 characters per 2.5 cm.
3. No more than 6 lines of type within a vertical space of 2.5 cm.
4. Margins, in all directions, must be at least 2.5 cm.

Considering the strict page limits set for proposals, applicants should keep in
mind that using the smallest allowable letter size to squeeze more content into

a page will make reading that much more difficult, so they should exercise careful judgment. It is also helpful to use a serif font (such as Times New Roman), which flows easier visually. Failure to follow prescribed formats may cause a proposal to be returned without review, requiring preventable resubmission and wasting precious research time.

The following discussion of key elements of a proposal's description incorporates an extended example of a successful submission to the National Science Foundation (NSF). The funded proposal was submitted in 2001 by the biologists Steven L. Lima and William A. Mitchell, and is titled "Large-Scale Phenomena in Anti-Predator Behavior: On the Consequences of Putting Predators Back into Predator-Prey Interactions" (hereafter referred to as "Predators").[5] The behavioral system they studied was that of *Accipiter* hawks preying on such small birds as dark-eyed juncos (*Junco hyemalis*), geographically situated in and around Vigo County, Indiana. As is typical with government proposals, Lima and Mitchell's project was not initially funded; therefore, before resubmitting it they added elements that responded specifically to reader critiques.

PROPOSAL TITLE AND SUMMARY

The proposal's title and summary should not be considered mere formalities to be added after the project is fully described in the main text of the grant application. Given the prominent location and therefore immediate visual and cognitive impact of those two first items, applicants should not underestimate how much weight they carry with readers. An appealing title for a proposal may entice a busy reviewer to select that one to read ahead of others, and an effective summary may either reinforce a title's positive impression or cause the reviewer to have second thoughts. The title and proposal also will demonstrate from the outset how well applicants are able to express their ideas for both specialized and lay evaluators. Besides the extended illustration used in this chapter, many examples of titles and descriptions of funded proposals can be found on the Web sites of sponsoring agencies.

THE TITLE

As with an article manuscript, the title of a grant proposal may take various forms, such as declarative statements, descriptions, questions, or, like Lima and Mitchell's proposal, a subject identification followed by a subtitle indicating the activity and significance of the research. An effective title must capture the essence of the project's conceptual framework. In constructing a title that accurately represents the basic idea of the project, applicants must walk a fine line, giving careful thought to its clarity, conciseness, specificity, and overall choice of words, including selectivity in using jargon and professional tone. In identifying the subject of their project, Lima and Mitchell's title, "Large-Scale Phenomena in Anti-Predator Behavior," is specific without being overwrought. Placing the phrase "large-scale phenomena" at the beginning of the title draws attention cognitively and conceptually to the work's magnitude as it points to their specific area of concern, anti-predator behavior. The subtitle, "On the Consequences of Putting Predators Back into Predator-Prey Interactions," adds a dynamic dimension, speaking of the consequences of an action—putting predators back into a circumscribed activity that they will study, predator-prey interactions. Overall, the title gives readers a concise yet specific and dynamic picture—with a minimum of jargon—of the scientific concept and research activity that form the investigators' subject.

THE SUMMARY

Although the terms "summary" and "abstract" may be taken as synonymous, they are distinguishable in that the reduction of content in an abstract is nonlinear while that of a summary is linear.[6] An abstract may omit some parts of a document's text and reduce the included ones disproportionately, while a summary is all-inclusive and proportional. A summary is intended to give readers a nutshell rendition of a document, and often is longer than the more selective abstract. The NSF's instructions to grant applicants ask specifically for a "project summary" that focuses on the experimental activity (emphasis added).

Ex. 10.2

The proposal must contain a summary of the proposed activity suitable for publication, not more than one page in length. *It should not be an abstract* of the proposal, but rather a self-contained description of the activity that would result if the proposal were funded. The summary should be written in the third person and include a statement of objectives and methods to be employed. It must clearly address in separate statements (within the one-page summary): (1) the intellectual merit of the proposed activity; and (2) the broader impacts resulting from the proposed activity.

The three-paragraph summary for "Predators," shown in Ex. 10.3, adheres to the NSF guidelines and exhibits various qualities that make it maximally effective. First, the flow of the narrative has an evident logic, moving with progressive specificity from establishing the work's need and merit to posing the key research question and delineating the theoretical framework, to describing the experimental activities to be conducted. Second, the language is kept simple and readable for nonspecialized readers. The behavioral phenomena, theoretical constructs, experimental activities, and data collection all are described without occlusive technical jargon. Let us consider the specific features of the full summary for "Predators," which reads as follows:

Ex. 10.3

Most studies of anti-predator behavior focus on the scale at which individual predators interact with individual prey. This "small-scale" perspective has revealed much about predator-prey interactions, but has largely missed important behavioral interactions that exist at larger spatial scales. This situation is due in part to the fact that behavioral ecologists have largely ignored the strategic role of predator behavior in predator-prey interactions. To appreciate large-scale interactions, one must not only understand the nature of predator behavior (especially large-scale movements), but also think on an ecological scale much larger than usually considered by behavioral ecologists. Large-scale interactions occur, in essence, because the movement of a predator acts to link the behavior of prey across the predator's home range, even if the prey in question do not directly interact with one another. This indirect interaction among prey, in turn, ensures that the predator will be on the move. The objective of the proposed work is to explore the nature of these large-scale behavioral interactions in an avian system. Such work represents an important step in the behavioral ecology of predator-prey interactions, one that may ultimately shed much light on predator-prey interactions in general.

The proposed work is motivated by a simple question: what drives the movement of prey animals across a landscape? The present consensus holds that movement attracts predators, hence movement by prey must reflect a need to find new sources of food, mates, etc. However, major prey movements may be a form of predator avoidance—and a manifestation of large-scale predator-prey interactions. The results of a simulation model show that predator and prey may be involved in a large-scale *"shell game,"* in which predators search for elusive prey, and prey move among feeding sites to re-

---

main elusive. The crux of the shell game is that the risk of attack experienced at a given site is directly related to the degree to which prey are predictably at that site. Furthermore, if prey in a given area are relatively difficult to locate, then predators will focus their attention elsewhere on more predictable prey, with the result that these "predictable" prey may initiate greater movement to avoid predators. This indirect interaction among prey represents another large-scale phenomenon: *the predator pass-along effect*. On the other hand, prey may choose to predictably bias their feeding toward certain very profitable feeding patches. Under these circumstances, predators might avoid constantly attacking prey in an effort to render them more catchable in the long run; this spreading of risk across a predator's home range is another large-scale phenomenon—*prey management by predators*. The further theoretical exploration of these large-scale phenomena is an important feature of the proposed work.

The bulk of the proposed work concerns an empirical exploration of these large-scale behavioral interactions. This work focuses on the conceptually important paradigm of the "small bird in winter," in which small birds like dark-eyed juncos (*Junco hyemalis*) must survive the rigors of winter and the predatory onslaught of bird-eating *Accipiter* hawks. Applying these ideas to the small-bird-in-winter paradigm will require a good deal of basic biological research, as very little is known about the large-scale movements of wintering birds, especially *Accipiter* hawks. The daily movements of such hawks and their avian prey (flocks of juncos) will be followed (via radio-tracking techniques) over study areas many km^2 in extent. Basic information will be established on hawk and prey home range sizes and movements. Using food manipulation sites established throughout a study area, the distribution and spatial predictability of prey will be altered experimentally. These food sites will also allow for the testing of a basic assumption underlying the shell game—that hawks bias their activity toward areas where prey are predictable. The removal of a portion of the food sites in a study area will be used to determine whether the resulting (simulated) predator pass-along effect leads to the increased movement of flocks not directly affected by food site removals. Food sites will also allow for a determination of whether hawks refrain to some extent from attacking birds at predictable feeding sites as suggested by computer simulations of prey management. Furthermore, the close tracking of hawks will enable the study of small-scale hawk-flock interactions that may drive certain large-scale interactions. Conducting behavioral research at large spatial scales presents unique challenges, but the insights gained will be worth the effort.

A close reading of the "Predators" summary reveals that each paragraph serves a clear function that is connected logically and clearly to the two other paragraphs. Appropriately, the first paragraph begins by justifying the study's focus on large-scale behavior—reflected in the initial words of the proposal's title—by emphasizing that studies at smaller spatial scales miss important behaviors. The authors then define large-scale interactions as a linkage of predator and prey movement, and state their objective to study such behavior within "an avian system." The final sentence points to the broader significance of the proposed research as "an important step in the behavioral ecology of predator-prey interactions, one that may ultimately shed much light on predator-prey interactions in general." The second paragraph immediately states the key question for the proposed work: "What drives the movement of prey animals across a landscape?" To find an answer, the applicants propose to study three large-scale behavioral phenomena, conceptualized in simple terms as a "shell game," a "predator pass-along effect," and "prey management by predators." In the third paragraph, the authors state the project's focal paradigm ("small bird in winter"), identify the avian species involved (*Junco* prey, *Accipiter* hawks), and describe their experimental activities (radiotracking, food manipulation).

Although no formula exists for devising the most effective summary, the biologists Andrew Friedland and Carol Folt observed that "some of the most compelling summaries start with a broad statement of purpose and then funnel the reader to the specifics of the proposed work."[7] The effectiveness of the "Predators" summary is evident in its funneling approach combined with its straightforward language used to convey sophisticated behavioral dynamics in the field.

We can turn now to the basic elements of a project's full description, which the NSF limits to 15 pages. The 15-page project description for "Predators" comprises the following numbered primary headings:

Ex. 10.4

1. Overview of Changes to Proposal
2. Introduction and Proposal Overview
3. Results from Prior Support
4. Background Information
5. Application to the Small-Bird-in-Winter Paradigm
6. Preliminary Work
7. Proposed Work
8. Broader Impacts of the Proposed Research

This chapter takes up each of the elements that typify a fully described project, as exemplified by these section headings in the "Predators" proposal. We will begin with the component that addresses changes in resubmitted proposals made in response to reviewer critiques.

REVISIONS IN RESUBMITTED PROPOSALS

Most successful proposals are submitted more than once to their sponsoring agency before finally being funded, so a common part of grant applications is a direct response to reviewer critiques in which applicants describe the nature of their revisions to an earlier version. The importance of how a resubmitted proposal responds to the concerns of reviewers—fully, specifically, and concisely—cannot be overstated if a second hearing is to work in its favor. In "Predators," a one-page "Overview of Changes to Proposal" is prominently placed before the introductory section, a sound practice that makes it easier for busy readers to quickly grasp the essence of a work that has been reviewed previously. The overview responds specifically yet succinctly to the five areas of concern that were raised by reviewers, devoting to each one a single paragraph (summarized here).

- *Panel Issue 1: Flock stability and movement.* Reviewers questioned the assumption that prey flocks "are stable over time and move as a cohesive unit"; the authors describe their efforts to answer the issue by collecting pilot data—given in the Preliminary Work section—which also supported estimates of prey home ranges and feeding patterns.
- *Panel Issue 2: Other predators.* Concerns about the mixed presence at study sites of unmarked and marked or tracked hawks are addressed by assurances that unmarked predators would be detected and accounted for in the data. Additional predators, such as other hawks, coyotes, foxes, and cats, are characterized as posing only a minor threat to the prey in the study.
- *Panel Issue 3: Factors other than predation (winter weather).* The issue of potentially disruptive effects of variable winter temperatures is addressed by mention of pilot work showing such effects to be minimal. The issue of heavy snow cover affecting feeding patterns at experimental sites, and thus disrupting the predator-prey shell game, is addressed by a protocol change "to equalize the number of low-snow/bare ground days across control and manipulating treatments."

- *Other Issues.* Here the authors take up suggestions made by individual reviewers: Should the study be scaled back in favor of more intensive tracking of *fewer* hawks? Do prey flocks alter their feeding patterns based on changing resources? Can baseline data be collected for prey movement in the absence of predators? The authors respond that, respectively, a larger study is better, feeding resources are stable, and field data for baseline prey movement is unobtainable.
- *New Pilot Work.* Pilot work on tracking predators, completed since the initial proposal, is offered to verify hawk behavior patterns and to show improved radiotracking methods that yield better data.

The overview of changes to the initial proposal is an effective response to the critiques of the review panel as a whole (first three paragraphs) as well as those of the individual reviewers (fourth paragraph). Another strength is that any countering views to particular critiques are phrased in positive and conciliatory language (italicized here): "This assumption *seems reasonable,* but . . ."; "tracking data from days with heavy snow cover . . . *may provide valuable insights*"; "tracking many hawks would be better. . . . *Nevertheless, we will be able to adopt* . . ."; and, "*We can indeed see the value* in the intensive tracking of hawks. However, . . ." Such phrasing shows that the critiques have been taken seriously and with an open mind, rather than dismissively, which also minimizes the potential risk of offending reviewers by making them feel they have wasted their time and effort in offering constructive comments. One cannot overemphasize the importance of both content and wording in describing the revisions made to an earlier version of a proposal in direct response to the concerns and suggestions of reviewers.

REFERENCES IN A PROPOSAL

Whatever extent of repetition or elaboration is necessary in the full text of a proposal beyond the summary and description of revisions, as the complete details emerge we begin to see citation of the literature. Generally, there is no set limit to the number of citations allowed. The 15-page project description for "Predators" is followed by a 9-page list of references that includes the applicants' own publications. As in a manuscript for a journal article, however, sources must be chosen selectively. Questions may arise as to how many citations should be included, which ones, for what purpose, and in which manner.

The literature must be reviewed thoroughly, but quantity is less important than quality. Are representative and key sources cited? Are they mostly recent? Are sources included that do not agree with the proposal's arguments or theoretical perspectives? Is there unnecessary citation due to tangential and thereby distracting information? Is every citation worth its weight for a direct, concise, and focused narrative of the purpose at hand? Another issue besides selectivity is the manner of citation. First, internal citation and the end-of-text listing of references must not only be accurate and complete but also use a consistent and conventional style. Some agencies require a specific format (such as CBE), although the following guidelines for NSF applicants are relatively open.

Ex. 10.5

Reference information is required. Each reference must include the names of all authors (in the same sequence in which they appear in the publication), the article and journal title, book title, volume number, page numbers, and year of publication. If the document is available electronically, the Website address also should be identified. Proposers must be especially careful to follow accepted scholarly practices in providing citations for source materials relied upon when preparing any section of the proposal.

Other than the inclusion of all identifying items in the reference, NSF applicants need only follow "accepted scholarly practices." The reference format in "Predators" resembles APA style but is actually a combination of accepted practices, as in this example:

Lima, S.L., and Zollner, P.A. (1996). Anti-predatory vigilance and the limits to collective detection: visual and spatial separation between foragers. *Behav. Ecol. Sociobiol.* **38**, 355–363.

A second point concerns the placement of citations in the flow of the narrative. Poorly positioned citations can confuse readers as to those works' direct relevance to each point being made. For instance, placing all citations at the end of a paragraph, rather than with the individual sentences that make the particular point to which they relate, will jeopardize a reader's sense of the precise connections to the literature. Moreover, the intended purpose of parenthetical citations will be more evident when they are preceded by precise use of such abbreviations or phrases as "e.g." (such as), "i.e." (that is), "see also," or "but see." Finally, as in other research documents, applicants should include only those references that they have actually read and digested. Simply copying widely used references from other publications runs the risk of including a citation that is marginally or not at all relevant to the point being supported.

INTRODUCTION AND BACKGROUND

An effective project summary, followed by a cogent response to critiques of a prior version of the proposal, should drive readers forward with interest as they review the complete details. The proposal's introductory and background information lays the groundwork for readers to comprehend the essence of any theoretical perspectives and experimental activities to be more fully explained as the narrative unfolds. What are the key hypotheses and theoretical concepts that are central to the project? What are the experimental goals? Has any preliminary work been done? Are there results from prior funding? How is the overall project idea significant? In "Predators," answers to these questions are provided over various sections that make up the project's full description, funneling toward a direct presentation of the experimental activities involved in the proposed work itself. Let us take a closer look at how these sections work together to help develop a detailed picture of what the investigators propose to do.

INTRODUCTION: SIGNIFICANCE, OBJECTIVES, AND HYPOTHESES

In their second section, "Introduction and Proposal Overview," the authors of "Predators" cite studies that establish the fundamental concept that all animals can engage in "decision making" in response to changes in the predatory situation. After introducing that framing concept, they underscore and elaborate upon their primary objective and two main issues, using keywords that readers can readily correlate with the proposal's title:

- Our main goal is to extend this work by exploring anti-predatory behavioral phenomena that occur at large spatial scales.
- First, we need to think about anti-predatory behavior on a scale much larger than the laboratory or a single field study site.
- Second, we need to put predators back into behavioral predator-prey interactions.

As the authors elaborate, they emphasize the limitations of existing knowledge and theories on predator-prey interactions (e.g., "few studies explicitly recognize," "very little is known about," "virtually nothing is known about"). They point specifically to the inadequacy of a current experimental paradigm that treats predator behavior as a "black box" abstraction exemplified by the use of "a fixed attack rate in many theoretical models." Their own hypothesis is stated in simple and now-familiar terms: "We believe that a key to understanding large-scale behavioral interactions is a much better understanding of large-scale predator movements, and how such movements respond to (and influence) decision-making by prey." Finally, they cite several studies to support the promise of the alternative conceptual paradigm of the "small-bird-in-winter," which is at the core of their proposed research.

BACKGROUND INFORMATION: REVIEWING PAST WORK AND PRESENT CONTEXT

After introducing the key concepts, objectives, and hypothesis for the proposed work, a proposal must provide contextual information that funnels readers toward a detailed description of their proposed experimental activities. "Predators" contains various kinds of such contextual information that collectively can be called "background." First, the applicants briefly summarize their results from prior NSF funding. Although that work is not directly related to the present proposal, the references list 14 articles that were published as a result of the funded project, which demonstrates a track record of success. Second, "Predators" contains a main background section that reviews the relevant literature. Finally, in subsequent background sections the applicants apply their literature review to the theoretical paradigm of their proposed work and then describe their preliminary experimental activities and results.

The literature review in a proposal should give the reader a general picture of the status of the problem. Rather than a comprehensive literature survey, what is required here is a brief, critical synopsis of those hypotheses, approaches, and results that constitute the present state of knowledge in the area under consideration. One is expected to include a balanced treatment of the work of all major investigators, not just one's own views and results. Nevertheless, applicants may also convey a sense of their own contributions to the field. The background information section in "Predators" reviews past work and addresses the key concepts and behaviors to be studied in answering the proposal's basic question regarding predator-prey interactions on a large scale: "Why do animals move?" The section is organized by subheadings, making it easily readable in outline form:

Ex. 10.6
4. Background Information
   A. Past Work
      1. Spatial scale in the study of anti-predator behavior
      2. Strategic interactions between predator and prey
   B. Large-Scale Interactions: The Importance of Predator Movement
      1. Predator-prey "shell games"
      2. Linking prey behavior across a landscape: the predator pass-along effect
      3. Predator management of prey behavior

In reviewing the published work, the applicants identify theoretical limitations in existing studies while arguing for the promise of focusing on the three large-scale behavioral phenomena in their proposal (4.B.1–3 in the example above). The background on these phenomena is then applied, once again in a funneling fashion, to the specific theoretical paradigm of the proposed work, the small-bird-in-winter. Like the preceding background section, the readability of this central theoretical component is enhanced by subsectioning:

Ex. 10.7
5. Application to the Small-Bird-in-Winter Paradigm
   A. Accipiter Hawks
   B. Small Wintering Birds: Large-Scale Movements
   C. Small Wintering Birds: Small-Scale Considerations
      1. Vigilance
      2. Behavior following encounters with predators
      3. Temporal patterns in risk

As the applicants highlight the limitations in existing studies, they underscore how their own objectives—within the small-bird-in-winter paradigm—promise to address gaps in the current understanding of the predator-prey behavioral phenomena in question. This overall strategy is reflected in the phrasing of their objectives: "Very little is known about . . . Accordingly, one of our main goals is to"; "Another of our main goals is to . . . Here too there is little available information"; or, "This information will also add a new dimension to." Such wording constantly reminds readers of the significant contributions offered by the proposed work and how those contributions will address the documented deficiencies in current knowledge on the subject.

PRELIMINARY WORK

Another supporting component of proposals provides background information beyond a literature review: a description of the applicants' preliminary experimental activities that are directly associated with their proposed project. Such information leads directly to a detailed picture of the proposed work. In the case of "Predators," the applicants describe their results and applicable conclusions from computer modeling and pilot fieldwork, organized by the following subheadings:

Ex. 10.8

6. Preliminary Work
    A. Modeling Large-Scale Behavioral Interactions
       1. A simulation model
       2. Simulation results
    B. Movements and Behavior of Hawks
       1. Trapping and radiotracking
       2. Home ranges and general movements
       3. Temporal patterns in activity
       4. Prey attacked and related behavior
    C. Movements of Junco Flocks
    D. Implications for the Proposed Work

As expected, at this point the details become very specific because the methodology is now directly associated with the work for which the applicants seek funding. The proposal as a whole has been funneling toward progressively greater specificity in describing what the project will entail. A precise and clear presentation of such preliminary work is a compelling factor in the judgment of readers as to the likelihood that the proposed project will be completed with timeliness, success, and significant outcomes. The high degree of specificity in the methodological details of the applicants' preliminary work is evident in the following excerpt that describes their computer modeling (from 6.A.1, "A simulation model," in the example above).

> Ex. 10.9
>
> We used computer simulations to investigate the validity of the large-scale phenomena outlined above. Our simulations were developed specifically for hawks and flocks of birds occupying a patchy landscape. Each simulated landscape consisted of 25 "patches" (potential prey feeding locations) and contained 2 hawks and 3 prey flocks, the latter with 20 birds each at the start of a simulation. A flock moved as a cohesive unit among 3 contiguous patches of non-depleting food, with one path yielding twice the energy intake of the other two.

In providing the results of their simulations, the applicants introduce the first of four figures used in the proposal—a bar graph "illustrating the predator-prey shell game and related large-scale phenomena." Two other figures, both maps, are used as the applicants describe how and where they tested methods for radiotracking and capturing hawks as well as followed the daily movements of junco flocks. The figures are numbered sequentially, labeled and keyed clearly, and captioned with several specific but concise explanatory sentences.

DESIGN AND METHODOLOGY OF THE PROPOSED WORK

Now we arrive at the most critical and detailed component of the proposal, which all the earlier sections have been leading toward: a complete and precise plan for what the applicants will do if their project is funded. This should be the longest component of a proposal, sparing no detail that will impress upon readers the resourcefulness and authority of the investigators in mapping out a plan that is sensible and doable given the described activities, resources, and time frame. Readers must derive a cogent sense of the specific methods, anticipated outcomes, and treatment of data for the proposed work. Techniques or procedures must be seen as suitable for their tasks (as well as safe and ethical) so that reviewers will not question their appropriateness in favor of alternative methods. (In "Predators" the applicants had already addressed such issues at the beginning, in the overview of changes in response to the critiques of their initial submission.) The experimental narrative of a proposal should also address the overall methodology as well as the expected outcomes and treatment of data.

Ex. 10.10
7. Proposed Work
   A. General Methods
      1. The system: predator and prey
      2. Study area and general issues of research design
      3. Food manipulations
      4. Determining prey availability within study areas
      5. Predator and prey movement
         Predator tracking
         Prey tracking
   B. Large-Scale Expectations and Analyses
      1. Predator-prey shell game
         Predator
         Prey
      2. Predator pass-along effect
         Predator
         Prey
      3. Management of prey behavior by predators
   C. Smaller-Scale Issues and Analyses
      1. Vigilance
      2. Post-attack dynamics
      3. Temporal patterns in risk
   D. Further Development of Theory

Under the description of their general methods (in 7.A.1–5), the applicants begin by reaffirming and justifying their selected animal subsystem—the dyad of sharp-shinned hawks (predators) and dark-eyed juncos (prey). As their narrative progresses, they provide precise information on the following basic aspects:

- *study area:* field sites in farmland and forest (with reference to map figures) over a 35-square-kilometer area of Vigo County, Indiana;
- *time frame:* requirement of three winter seasons, November through March;
- *data collection:* locations, procedures, and timing of data collection on predator and prey behaviors, including food manipulation, patterns of movement, feeding, and attack, as well as trapping and radiotagging;
- *analyses:* treatment of data—daily travel paths, 30-day observation blocks— to test behavioral assumptions and outcomes expected.

The methodology is supported by practical details that demonstrate its feasibility and by citation of studies that provide further details and alternative approaches. For instance, in describing how they will analyze one of their hypothesized phenomena—the shell game—the applicants exhibit authority, thoroughness, and pragmatism.

Ex. 10.11

A predator-prey shell game is based on the assumption that areas of predictable prey activity will be focal points of predator activity. We will assess this assumption by relating biases in hawk home range use to measures of prey abundance and prey predictability (as defined earlier) derived from point counts (which also cover food sites and home feeders). This can be done by pooling hawk locational fixes and point count data into discrete cells across a given home range (Neu et al. 1974; Dasgupta and Allredge 2000). An alternative analysis (Powell et al. 1997) that might prove useful is based on the point count data and a hawk's home range utilization density as determined by a kernal home range estimator (Worton 1989).

This excerpt on methodology represents the use of various effective strategies, particularly in restating a definition (first sentence), stating directly what will be done ("We will assess"), cross-referencing ("as defined earlier"), specificity ("This can be done by"), and showing awareness of options ("An alternative analysis that might prove useful"). In their description of expected outcomes, the applicants also demonstrate flexibility and adaptability regarding the analysis of unexpected results, as in the following sentence: "If neither of these outcomes is supported, then we will use a power analysis to determine how certain we can be that daily path use is indeed random." There is also a sense here of preemptive anticipation of a potential reviewer concern, which will tend to work in any proposal's favor.

Beyond such qualities of logical flow, clear and simple wording, and specificity in key details, the overall description of the proposed activities and methods illustrates other basic strategies that make the narrative readable and appealing. First, the writers do not assume that readers are familiar with their methods and therefore provide sufficient detail to visualize and evaluate their suitability. Second, the description of methods is still kept brief with the help of citations that contain the full details of procedures and techniques, such as those for collecting or analyzing data. Third, visuals are used effectively to illustrate experimental settings, methods, and sample results. In their description of how they will follow hawk movements, for instance, the applicants include a map figure with a caption that reads: "An example of a single-day travel path for sharp-shinned hawk 565 based on continuous tracking (red dots and lines) and the proposed method of locational fixes every 2 hours (open black circles, yellow lines)." Finally, the applicants are specific about timeline practicalities, both for the project as a whole and for periods or cycles of experimental manipulation, observation, and data collection. This allows readers to see that time frames are feasible and realistic, and that the project can indeed be carried out within the time allotted.

Having provided the details of what they will do experimentally, the applicants conclude their methodology section by describing how they will engage in "further development of theory" (subheading 7.D in Ex. 10.10). This is followed by the final section of the proposal, "Broader Impacts of the Proposed Research" (section 8 in Ex. 10.4). Both of these sections are important in providing reviewers with a specific sense of the upshot and ultimate contribution of the proposed project. The researchers' goal in theory development is to achieve a full understanding of the large-scale behavioral interactions between predators and prey, particularly the shell-game concept. As in the preceding sections, the discussion is specific and identifies various areas that are insufficiently understood, such as the behavior, movement, and energetic state of prey relative to feeding patches. The applicants' specificity is supported by citations that pinpoint how they expect to extend existing theory, as in the following sentence:

> Ex. 10.12
> We are particularly interested in integrating our modeling approach (to large-scale issues in habitat use) with the ideal-free based approach of Hugie and Dill (1994) and Sih (1998); this integration will likely revolve around assumptions concerning the independent movement of animals between patches and the degree of "spatial omniscience" assumed for predator and prey.

Beyond describing expected contributions to the theoretical understanding of its subject, a proposal should address what the project's success will mean more broadly, typically in education and in the public sector. The concluding paragraphs in "Predators" focus on benefits of the proposed work to students and to the community. Besides training for graduate students, the applicants offer to engage undergraduates in the "research experience," including tracking animals, using GIS and GPS technologies, and using software for theoretical modeling. To encourage public interest, they will promote environmental awareness by having a local nature center post project information on its Web site and offer hands-on learning about bird tracking to local school students, as well as by hosting hawk watches with the Audubon Society. Given that government funding for scientific research is derived from a taxed citizenry, specifying the return benefits to students and the public at large should not be considered merely an afterthought.

BUDGET PREPARATION

A final component required in grant proposals is a budget section that shows and justifies how the requested funds will be spent. Preparing the details of a budget calls for close attention to various elements, including project needs and their costs, institutional policies and protocol, and financial ethics. The budgetary items include *direct* costs for doing the research as well as *indirect* costs, or overhead expenses, which are calculated with the assistance of the institution's office of research and grants. In preparing a budget, applicants therefore should consult with their campus grants officers as well as the guidelines for itemizing provided by the granting agency. A proposal's budget typically includes costs for the following kinds of items:

- *salaries and stipends:* compensation for such personnel as principal investigators, graduate or undergraduate assistants, postdoctoral fellows, technicians, secretaries, consultants, and collaborators;
- *equipment:* deciding factors include need and rationale for request, expense relative to choice of make or model, lease-buy options, institutional matching funds, and overall cost;
- *materials and supplies:* categories and unit costs for such needs as animals, tissue culture, radioisotopes, histology, microscopy, photography, glassware, and computer hardware, software, or use time;
- *travel:* mileage, fare, lodging, food, and number of trips for such purposes as fieldwork, data collection abroad, conference presentations, and use of specialized equipment, facilities, or services;
- *miscellaneous:* smaller ancillary expenses such as for phone, photocopying, postage, and publication.

Table 10.1 Budget for "Predators," showing how the total funding request ($112,203) is divided into equal parts over the three years of the proposed project

| | Itemization Year | | | |
Line items	Year 1	Year 2	Year 3	Cumulative
Salaries (technicians)	13,500	13,500	13,500	40,500
Fringe benefits	270	270	270	810
Equipment	0	0	0	0
Travel	3,328	3,328	3,328	9,984
Materials and supplies	12,680	12,680	12,680	38,040
Total direct cost	29,778	29,778	29,778	89,334
Total indirect cost	7,623	7,623	7,623	22,869
Total request	37,401	37,401	37,401	112,203

Courtesy of Steven L. Lima, Indiana State University

Applicants should consult the granting agency guidelines for definitions and constraints of permissible budget items. For instance, NSF guidelines specifically define equipment as "an item of property that has an acquisition cost of $5,000 or more (unless the organization has established lower levels) and an expected service life of more than one year." Grant monies for compensation cannot be used to augment the salaries of regular faculty, while salaries for eligible personnel typically are calculated in "full-time-equivalent person-months." Budget restrictions also may apply to expendable materials or to foreign travel. All requested items must be fully justifiable, with costs that are estimated accurately and truthfully.

In "Predators," the applicants requested a total of $112,203, divided into equal amounts for each of the three years of the project's duration (Table 10.1). Some granting agencies, including the NSF, require a separate page for budget justification. This provides a further opportunity for applicants to explain their needs in direct and simple terms and to minimize potential doubts regarding particular needs that ultimately could become a decisive factor for reviewers in their level of support for the proposal. The justification statement also can explain the nature of institutional cost sharing, the availability of other funds or resources, and any strategic use of time, personnel, or equipment that collectively permit a lower and more prudent funding request. The justification page for "Predators" first notes an increase in the budget request (from initial submission) due to a change in institutional calculation of indirect costs and higher estimates for transmitters and radiotracking personnel. Then applicants provided the following rationales for their specific needs:

- *personnel:* stipends for three field workers who are not already paid as graduate assistants ($4,500 per worker per year plus 2% fringe benefits);
- *travel:* automotive fuel ($0.26 per mile) for five field workers who will travel 15–20 miles a day, or 12,800 miles per field season ($3,328/year);
- *materials and supplies:* corn meal for food manipulation sites ($1,200 per year) and radio-transmitters for juncos and *Accipiter* hawks (82 total at $140 each, or $11,480 per year), totaling $12,680 per year;
- *indirect costs:* assessed at 25.6% of total modified direct costs, that is, total direct costs minus equipment (single items greater than $2,500) and participant costs.

Other items that may need to be included and justified are consultant services (e.g., statisticians, epidemiologists) and work subcontracted at other institutions. Applicants should prepare their budget realistically and with accurate calculation of costs, neither underestimating (a risk for initiates) nor overestimating. A helpful start is to consult with colleagues or advisers who are seasoned grantees. A budget that is appropriately derived for its purposes, carefully balanced, and ethically beyond reproach will demonstrate competency overall and be more competitive.

THE CHALLENGE AND RESPONSIBILITY OF GRANT PREPARATION

In today's research environment, it is commonplace for scientists to seek external financial support for their projects. Moreover, success in acquiring grants is used as a major criterion by university departments and administrators in their annual evaluation of science faculty. The importance of acquiring such support is reflected in an array of grant-writing courses, workshops, toolkits, and online tutorials for science graduate students and faculty, widely offered by departments and research offices. A commercial industry of specialists also has cropped up that offers grant training and consulting services to individuals and organizations. The preparation of grant applications can be a tedious and trying process even for the most organized and experienced proposal writers. Notwithstanding the availability of considerable assistance, full responsibility for a proposal's quality rests squarely on the shoulders of the principal investigator. For instance, although the investigators write the proposal, it will be routed for mailing by a university's research office after the institutional support documents are attached. Before that office actually mails it, the authors should review the entire application package to ensure that it is it completely in order. A final point is that fully meeting the promise of the proposed work goes beyond completing the experimental work itself. There may be a requirement of progress or follow-up reports to the granting agency, and there is an assumed obligation to eventually disseminate the results through some form of publication. In the end, writing a grant proposal and experiencing its review process is a practical necessity as well as an important part of the professional self-assessment and growth of contemporary scientific researchers.

NOTES

CHAPTER 1. SCIENTIFIC ENGLISH

1. Bloomfield 1939, 1.
2. Locke 1992, 27.
3. Bronowski 1978, 49.
4. Glass 1965, 1259.
5. Spedding et al. 1876–1883; Sprat 1958, 133 (emphases in original); and Jones 1951, 157. Though Sprat refers to "speaking" plainly, the Royal Society's linguistic strictures also applied to writing, as shown for instance by Adolph 1968.
6. Adolph 1968, 168.
7. Willey 1934, 214.
8. The passages quoted here are from Redish 1985, 125 and 129, respectively.
9. Day 1995, x (emphases in original). Commentaries and theories on plain writing like those in the mid-twentieth century—such as by Chase (1953) and Flesch (1949)—are now having their day, for example in the form of grammar checkers in word-processing software.
10. These steps are adapted from Hatch 1983, 44–45, which relied on Gunning 1968, 38–39.
11. Medawar 1979, 3 (emphasis in original); Luria 1984, 159 and 160.
12. Halloran and Bradford 1984, 183; Niven 1890. Maxwell's support of scientific metaphor is from an address to the mathematical and physical sections of the British Association, first published in *British Association Report,* 1870, vol. 2.
13. Gould and Lewontin 1979. For various examples of experimentation in under-

graduate pedagogy with creative expression in technical and scientific writing, see Whitburn 1978.

14. Kuhn 1970.
15. See Gould 1980, 154.
16. See Bleier 1976, 34 and 44 (emphasis in original). In response to Edward O. Wilson's controversial theory of sociobiology in the 1970s see Gould 1977, as an argument for biological potential over biological determinism.
17. See the historical perspective on this biomedical issue in Brandt 1997.
18. Katz 1985, 15 and 16; Gopen and Swan 1990, 553. Later, the chapter looks at some detailed examples of this basic point provided by the authors. Gopen and Swan's collaboration is striking in that Gopen holds a PhD in English and a JD (both from Harvard), and directs the writing program at Duke, while Swan earned a PhD in biochemistry (MIT) and teaches scientific writing at Princeton.
19. Freedman 1958; examples that are not cited are original, including some from my own experience in alcohol research.
20. Katz 1985, 16. Exs. 1.2 and 1.3 are from, respectively, Watson and Crick 1953, 737, and Jaffee et al. 2002, 262.
21. Jansen et al. 1994, 1227 (emphasis added).
22. American Psychological Association 2001, 38.
23. Perelman et al. 1998, 282–284, which discusses language that is "ageist," insensitive to disability, or ethnically or racially biased.
24. Wilkinson 1991, 63.
25. Fields 2004, 48.
26. Exs. 1.22 and 1.23 are from, respectively, Watson and Crick 1980, 264, and Gould and Lewontin 1979, 584 (for rhetorical analyses of this article's scientific prose, see Selzer 1993).
27. Medawar 1979, 63.
28. Council of Biology Editors 1994, 194.
29. Ex. 1.30 is from Freedman 1958, 12. Gopen and Swan, 1990, 550; their widely read article emphasizes the key areas of reader expectation illustrated here.
30. Gopen and Swan 1990, 552–553.
31. Exs. 1.32 and 1.33 are from ibid., 557; the surrounding text summarizes the method described in ibid., 552–557.
32. Ex. 1.35 and its revision in Ex. 1.36 are adapted from Goldbort et al. 1976.
33. Exs. 1.37 and 1.38, and surrounding text, are from Gopen and Swan 1990, 556.
34. The quotes are from Alley 1996, 119, and Wilkinson 1991, 67.
35. Alley 1996, 123.
36. Ex. 1.46 is from Matthews et al. 1996, 118; also see Medawar 1979, 63.
37. Day 1995, 128.
38. Katz 1985, 21.
39. McMillan 1997, 131.

40. Thomas 1979, 125–126.
41. Wilkinson 1991, 22; Day 1999, 99 (emphasis in original).
42. The four sentences—one of which, incidentally, contains a usage error that the editors did not catch—are quoted from, respectively, Schneider et al. 1972; Strange et al. 1976; Goldbort et al. 1976; and Tampier and Quintanilla 2002.
43. Ebel et al. 1990, 5 (emphases in original); Gopen and Swan 1990, 550; and Davis 1997, 2.

CHAPTER 2. LABORATORY NOTES

1. Ebel et al. 1990, 11.
2. Committee on Professional Training 2003, 10 and 12, respectively.
3. Snow 1934, 103.
4. Weaver et al. 1986. Nobel laureate David Baltimore did sign the retraction. The team member whose lab notes were questioned—and found to be sparse, messy, and altered—was Imanishi-Kari, who (along with Reis) did not sign the retraction. Some of the relevant documents in this case are assembled in Beall and Trimbur 1996.
5. Kanare 1985, 11.
6. Ibid., 15. A specific and official standard for permanence of writing paper—D3290-00 Standard Specification for Bond and Ledger Papers for Permanent Records, 2003—is available from the American Society for Testing and Materials (ASTM), in Philadelphia, PA.
7. The extended example in alcohol studies used here is adapted from my own research for an MS in biology, completed in 1975 at Indiana University of Pennsylvania and published in Goldbort et al. 1976, and Strange et al. 1976.
8. Ebel et al. 1990, 11.
9. Cornell Center for Technology, Enterprise, and Commercialization 2005.
10. See Beall and Trimbur 1996, 70, where the authors warn that instructions to "read and add but not change" data files can be circumvented, and point out that "in recent cases of alleged scientific dishonesty, computer records of laboratory work have been considered extremely weak evidence."

CHAPTER 3. WORKPLACE SCIENTIFIC WRITING: LETTERS, MEMORANDA, AND ABSTRACTS

1. Greenly 1993, 45–46.
2. This is an actual reprint request I received.
3. Adapted from actual correspondence with Professor John Bryan in the English Department at the University of Cincinnati, who replied on March 27 by placing his answers between the questions.
4. Quoted from form letters written by Sagan and Kendall.
5. Crabbe et al. 1994, 1715.
6. Goldbort 1975.

7. Goldbort and Hartline 1975, 720. I presented this research on April 14, 1975, at the Federation of American Societies for Experimental Biology, 59th Annual Meeting, Atlantic City, NJ, held April 13–18.
8. Goldbort et al. 1976, 263
9. Walker 1996, 39.
10. Day 1988, 30.
11. Wilkinson 1991, 359; Montgomery 2003, 83; Alley 1996, 20.

CHAPTER 4. UNDERGRADUATE REPORTS IN THE SCIENCES

1. Courses and programs in scientific writing burgeoned in the 1980s. See for instance Verbit 1983. Undergraduate and graduate courses in scientific writing, easily found online, currently are offered at such institutions as Clemson University, Pennsylvania State University, Princeton University, Rutgers University, University of California, Santa Barbara, University of the Pacific, and University of Arizona. As an example of a cross-disciplinary approach, both biology and English students also could read physician Robin Cook's high-tech re-creation of Mary Shelley's monster in his 1989 novel *Mutation,* the protagonist of which is geneticist Victor Frank, together with the illuminating 1986 essay by biologist Leonard Isaacs that compares Shelley's scientist with both the "father of the atomic bomb," J. R. Oppenheimer, and the creators of genetic engineering.
2. This table is partly adapted from Wilkinson 1991, 83.
3. National Academy of Sciences 1995.
4. Watson 1968, 3.
5. See Barnum and Carliner 1993, 111–127, for an extensive discussion of how personality differences can be used to assign team roles.
6. See for instance Brown and Keeley 2001.
7. A famous example is Watson and Crick's 1953 short announcement in *Nature* of their Nobel-winning elucidation of DNA's structure, a now almost larger-than-life article.
8. Fleischmann and Pons 1989.
9. See Djerassi's case study of scientific trust in his novel *Cantor's Dilemma.* An accomplished researcher, Djerassi uniquely aims to teach scientific concepts through his series of novels and plays (listed at www.djerassi.com).
10. From the print issue of October 22, 2004, 306 (5696), 557–760.
11. These are results from actual searches done on October 12, 2003.
12. Wilkinson 1991, 69.
13. Ibid.
14. Goldbort et al. 1976, 263–264.
15. Alley 1996, 252–253.

CHAPTER 5. DOCUMENTATION OF SCIENTIFIC SOURCES

1. National Academy of Sciences 1995, 12. For simplicity, use of the word "paper" here is intended as inclusive of college reports and journal articles.
2. These five manuals are, respectively, (1) Council of Biology Editors 1994 (repr. 1997), which uses a citation-sequence system (illustrated here) and a name-year system, both based on the 1991 *National Library of Medicine Recommended Formats for Bibliographic Citation;* (2) Dodd 1997; (3) American Psychological Association 2001; (4) *Chicago Manual of Style* 2003; and Gibaldi 2003.
3. Standardized abbreviations for journal titles are available from such organizations as the National Information Standards Organization (NISO) in Bethesda, MD, and the American National Standards Institute (ANSI) in Washington, DC. For chemists, see Dodd 1997, 215–229, which lists abbreviations for more than a thousand of the most commonly cited journals.
4. For some examples of fledgling efforts to create specialized scientific databases through academic-corporate partnerships in the early 1990s, see Goldbort 1995.
5. When the 6th edition of the CBE manual appeared in 1994, based on the 1991 guidelines of the National Library of Medicine (NLM), the Internet was in its infancy. In July 2001, NLM issued a supplement for Internet citation formats (compiled by Karen Patrias), which will be incorporated in a forthcoming 7th edition of CBE to be published by the Council of Science Editors (CBE's new name as of January 1, 2000). Therefore, the examples here follow NLM 2001.
6. NLM 2001, i.
7. The quotation is from Yamamoto 2003, S8.
8. Hall 1994, 1713 (References and Notes, entry 32). In contrast to many scientific journals, *Science* allows very lengthy notes.
9. The examples are from, respectively, (1) Galazka et al. 1999, 5 (boldface in original); (2) Montalto 2002, 344; and (3) Stanley and Yang 1994, 1342.
10. Instructions to Authors for *Physiological and Biochemical Zoology* are available online at http://www.journals.uchicago.edu/PBZ/instruct.html (retrieved May 25, 2004).
11. Montgomery 2003, 92–93.

CHAPTER 6. SCIENTIFIC VISUALS

1. See Djerassi 1994, chap. 19, and Crichton 1991, 161–165.
2. Briscoe 1996, 5; Montgomery 2003, 113–114.
3. Wilkinson 1991, 161.
4. Briscoe 1996, 41.
5. Tables 6.1 and 6.2 are from Ramstedt 2002, 310 and 314.

6. Table 6.3 is from Naimi et al. 2003, 73.

7. Nurnburger, Jr., et al. 2002, 235.

8. Council of Biology Editors 1994, 691.

9. Adapted from Pridemore 2002, 1922, 1924, and 1927.

10. Gratitude for Figures 6.3 and 6.4 goes to Michael Angilletta, who used them in his dissertation work (see Chapters 8 and 9).

11. Katz 1985, 53–54.

12. Dick and Foroud 2002, 173 (italics in caption added).

13. See Hall 1994, 1710, Figure 5, for an example of a caption having more than 700 words (for a schematic representing sex determination in the fruit fly, *Drosophila*).

14. Heath and Nelson 2002, 197.

15. Anni and Israel 2002, 222.

16. There are book-length treatments of scientific visuals that are considerably more comprehensive in scope and detail than is possible within a single chapter. Examples are Wolff and Yeager 1993; Briscoe 1996; Tufte 1997; and the 154-page chapter "Tables and Illustrations" in Wilkinson 1991.

CHAPTER 7. SCIENTIFIC PRESENTATIONS

1. See Glass 1988, 42. In his 1626 *New Atlantis* (published posthumously), a utopian dream-vision of a modern institute for experimental research, Bacon included publication resources for keeping the citizenry informed of scientific advances.

2. Medawar 1979, 61.

3. Ibid., 59.

4. Goldbort and Hartline 1975. I gave this presentation at the 59th meeting of the Federation of American Societies for Experimental Biology (FASEB), Atlantic City, NJ, April 14, 1975.

5. Goldbort et al. 1976.

6. Ebel et al. 1990, 343.

7. Goldbort and Cochran 2003 (personal photograph). The *Consensus Statement* was developed by the Indiana Perinatal Network's Subcommittee on Postpartum Depression, chaired by the poster's first author.

8. Both vertical and horizontal layouts are illustrated cogently by Matthews 1990.

9. Schowen 1997, 27–38.

CHAPTER 8. SCIENTIFIC DISSERTATIONS

1. For an example of a "Brief Communication" that lists two undergraduates as second authors, see Schneider et al. 1974.

2. Two online examples are Indiana University's *A Guide to the Preparation of*

Theses and Dissertations, http://www.indiana.edu/~grdschl/guide.html, and the University of Arizona's *Manual for Theses and Dissertations,* http://grad. admin.arizona.edu/degreecert/thesismanual/front.htm (both retrieved April 16, 2004). General guides include Fitzpatrick et al. 1998 and Bolker 1998). Personal narratives of the process across disciplines are offered in Pyrczak 2000. Examples of more formal and discipline-specific guides are Wilkinson 1991; Cone and Foster 1993 (repr. 2001); and Garson 2002.

3. Medawar 1979, 63 (emphasis in original); Woodford 1986, v, and see Woodford 1999, a thoroughly rewritten version of the teaching manual.
4. For examples of instruction available in science communication during the 1970s and 1980s, see Friedman et al. 1978, and Verbit 1983. Current courses and programs are now easily found online.
5. The alcohol project illustrated in Chapter 2 led to a 41-page MS thesis in biology.
6. Cooper 1986, 134–135.
7. Ibid., 136.
8. Angilletta currently is an assistant professor in life sciences at Indiana State University. He provided the copy of the dissertation from which the selected examples are taken. At the University of Pennsylvania, PhD students must follow the format prescribed in the *Doctoral Dissertation Manual* issued by Office of Graduate Studies, available online at http://www.upenn.edu/grad/DissManual. html.
9. Cooper 1986, 136.
10. A more comprehensive checklist of methodology items that includes clinical and qualitative research elements can be found in Cone and Foster 1993, 132–133.
11. Cooper 1986, 140.
12. Cone and Foster 1993, 222.
13. Cooper 1986, 141.
14. Garson 2002, 267. These elements originally were described by Toulmin et al. 1978.
15. Angilletta, Jr., 2001.

CHAPTER 9. SCIENTIFIC JOURNAL ARTICLES

1. Glass 1965, 1259.
2. Day 1988, 9.
3. Thanks to Michael Angilletta for our discussion that drew distinctions among the different types of theoretical constructs.
4. The *Journal of Nutrition*'s instructions for manuscript preparation are available at http://www.nutrition.org/misc/ifora2.shtml#prep (retrieved October 12, 2004).

5. "Manuscript Format and Organization" guidelines for the *Journal of Studies on Alcohol* are available at http://alcoholstudies.rutgers.edu/journal/infoforcont2. html (retrieved October 12, 2004).

6. The *Journal of Molecular and Cellular Biology* provides manuscript organization and format guidelines at http://mcb.asm.org/misc/itoa.pdf (retrieved October 12, 2004).

7. Mohanty et al. 1994, 427.

8. Author instructions for the *Journal of Virology* are available at http://jvi.asm. org/misc/itoa.pdf (retrieved October 12, 2004).

9. Wells et al. 2002, 319.

10. Angilletta, Jr., 2001, 19. Because this is an article based on a dissertation chapter, there is an institutional acknowledgement of the University of Pennsylvania. Subsequent page references will be made here to Angilletta's article. Personal appreciation is extended to the author for discussing his revisions based on peer reviews.

11. Instructions to authors for *Physiological and Biochemical Zoology* are available at http://www.journals.uchicago.edu/PBZ/instruct.html (retrieved October 14, 2004).

12. In a personal communication, Angilletta has noted that parenthetical references to "Angilletta 2002" in his article actually are inaccurate because the cited article, then in press, actually appeared sooner than expected, in November 2001.

13. Morimoto et al. 1995, 2954.

14. Katz 1985, 32.

15. The instructions to authors, revised in January 2005, are in *Biochemistry* 44, no. 1, 15A-20A, and are available at https://paragon.acs.org/paragon/ShowDoc Servlet?contentId=paragon/menu_content/authorchecklist/bi_authguide.pdf (retrieved August 19, 2005).

16. Hunter et al. 2003, 1155.

17. Information for contributors is posted by the *Journal of Chemical Physics* online at http://jcp.aip.org/jcp/submit.jsp (retrieved October 16, 2004).

18. National Academy of Sciences 1995; Pascal 2000, 222 (discusses federal and university policies on scientific misconduct, including ethics litigated federally); Dodd 1997, 417–423. The American Institute of Physics author guidelines are available at http://www.aip.org/pubservs/ethics.html, and those for the American Geophysical Union at http://www.agu.org/pubs/pubs_guidelines. html (retrieved October 19, 2004). For an example of a graduate course in research ethics, see Hoshiko 1993.

19. Luria 1984, 117.

20. See Chapter 1's discussion of Broca and Gould, as well as of Bleier's example

of sex-biased interpretations in human and animal studies. Bernstein 1978, 133–142, discusses the extreme case of the politically imposed Orwellian state of genetics "research" in Russia from 1934 to 1964, initiated by the fanaticism of biologist Trofim Denisovich Lysenko and fully supported by Stalin and Khrushchev. For a recent example of a fully documented case of fraud in a scientific paper by a physicist working in industry, see Lucent Technologies 2002; a major element of the misconduct was the manipulation of data and their misrepresentation in visuals, such as statistical plots. See also Mello and Brennan 2003.

21. This case is mentioned in National Academy of Sciences 1995. Bell, who now uses her married name, Burnell, received her PhD in radio astronomy at Cambridge University (where she discovered pulsars) and has received numerous awards from British and American scientific bodies. See also the US federal cases discussed in Pascal 2000.

22. One highly publicized allegation of fraud, mentioned in Chapter 2, was associated with an article by Weaver et al. (1986) that was retracted in 1991 by four of the six authors. Imanishi-Kari, who (along with Reis) did not sign the retraction, was suspected of fabricating data and altering her laboratory notes after the fact. See for instance Natalie Angier, "Rockefeller U. Anxious as Leader Feels Pain of Science-Fraud Case," *New York Times* (April 1, 1991), and Margot O'Toole, "The Whistle Blower and the Train Wreck," *New York Times* (April 12, 1991, op-ed page).

CHAPTER 10. SCIENTIFIC GRANT PROPOSALS

1. Woodford 1999, 104. See also the CBE's booklet on grant editing, Klein 1999.
2. Montgomery 2003, 148. For a comprehensive analysis of two biologists' grant proposals, see Myers 1990, 41–62.
3. See for instance the NSF's 61-page *Grant Proposal Guide* (NSF 04-23, July 2004), referred to in this chapter and available at www.NSF.gov, or the NIH's 107-page *DHHS Public Health Service Grant Application* (PHS 398, rev. 9/04), available at www.NIH.gov (both retrieved November 12, 2004).
4. Examples of recent guidebooks are Friedland and Folt 2000; Ogden and Goldberg 2002; and Blackburn 2003.
5. Lima, the principal investigator, received a PhD from SUNY-Binghamton and is professor of life sciences at Indiana State University; Mitchell was a postdoctoral research fellow at Indiana State University. Gratitude is extended to Professor Lima for permitting the use of his proposal as an extended illustration.
6. See Wilkinson 1991, 348, for a helpful discussion of these differences.
7. Friedland and Folt 2000, 63.

REFERENCES

Adolph, Robert. 1968. *The rise of modern prose style.* Cambridge, MA: MIT Press.

Alley, Michael. 1996. *The craft of scientific writing,* 3rd ed. New York: Springer-Verlag.

American Psychological Association. 2001. *Publication manual,* 5th ed. Washington, DC: American Psychological Association.

Angilletta, Jr., Michael J. 1998. Energetics of growth and body size in the lizard *Sceloporus undulatus:* Implications for geographic variation in life history. PhD dissertation, University of Pennsylvania.

———. 2001. Variation in metabolic rate between populations of a geographically widespread lizard. *Physiological and Biochemical Zoology* 74, no. 1: 11–21.

Anni, Helen, and Yedi Israel. 2002. Proteomics in alcohol research. *Alcohol Research and Health* 26, no. 3: 219–232.

Barnum, Carol M., and Saul Carliner. 1993. *Techniques for technical communicators.* New York: Macmillan.

Beall, Herbert, and John Trimbur. 1996. *A short guide to writing about chemistry.* New York: HarperCollins, 17–33.

Bernstein, Jeremy. 1978. *Experiencing science: Profiles in discovery.* New York: Basic.

Blackburn, Thomas R. 2003. *Getting science grants: Effective strategies for funding success.* San Francisco: John Wiley.

Bleier, Ruth. 1976. Myths of the biological inferiority of women: An exploration of the sociology of biological research. *The University of Michigan Papers in Women's Studies* 2, no. 2: 39–63.

Bloomfield, Leonard. 1939. *Linguistic aspects of science.* Chicago: University of Chicago Press.

Bolker, Joan. 1998. *Writing your dissertation in fifteen minutes a day.* New York: Henry Holt.

Brandt, Allan M. 1997. Just say no: Risk, behavior, and disease in twentieth-century America. In *Scientific authority and twentieth-century America,* ed. Ronald G. Walters. Baltimore: Johns Hopkins University Press, 82–98.

Briscoe, Mary Helen. 1996. *Preparing scientific illustrations,* 2nd ed. New York: Springer-Verlag.

Bronowski, Jacob. 1978. *Magic, science, and civilization.* New York: Columbia University Press.

Brown, M. Neil, and Stuart Keeley. 2001. *Asking the right questions: A guide to critical thinking,* 6th ed. Upper Saddle River, NJ: Prentice Hall.

Chase, S. 1953. *The power of words.* New York: Harcourt Brace Jovanovich.

Chicago manual of style. 2003. 15th ed. Chicago: University of Chicago Press.

Committee on Professional Training. 2003. *Undergraduate professional education in chemistry: Guidelines and evaluation procedures.* Washington, DC: American Chemical Society.

Cone, John D., and Sharon L. Foster. 1993 (9th printing, 2001). *Dissertations and theses from start to finish.* Washington, DC: American Psychological Association.

Cook, Robin. 1989. *Mutation.* New York: Putnam.

Cooper, Edwin L. 1986. Preparation for writing the doctoral thesis. In *Scientific writing for graduate students.* Bethesda, MD: Committee on Graduate Training in Scientific Writing, Council of Biology Editors, 134–142.

Cornell Center for Technology, Enterprise, and Commercialization (formerly Cornell Research Foundation). Guidelines for maintaining laboratory notes. Ithaca: Cornell University. http://www.cctec.cornell.edu.labnotebooks.html.

Council of Biology Editors. 1994 (repr. 1997). *Scientific style and format: The CBE manual for authors, editors, and publishers,* 6th ed. Cambridge: Cambridge University Press.

Crabbe, John C., John K. Belknap, and Kari C. Buck. 1994. Genetic and animal models of alcohol and drug abuse. *Science* 264, no. 5166: 1715–1723.

Crichton, Michael. 1991. *Jurassic park.* New York: Ballantine.

Davis, Martha. 1997. *Scientific papers and presentations.* New York: Academic Press.

Day, Peter. 1999. *The philosopher's tree: Michael Faraday's life and work in his own words.* Bristol, UK: Institute of Physics.

Day, Robert. 1988. *How to write and publish a scientific paper,* 3rd ed. Phoenix: Oryx.

———. 1995. *Scientific English: A guide for scientists and other professionals,* 2nd ed. Phoenix: Oryx, x.

Dick, Danielle M., and Tatiana Foroud. 2002. Genetic strategies to detect genes involved in alcoholism and alcohol-related traits. *Alcohol Research and Health* 26, no. 3: 172–180.

Djerassi, Carl. 1989. *Cantor's dilemma.* New York: Penguin.

———. 1994. *The Bourbaki gambit.* Athens: University of Georgia Press.

Dodd, Janet S., ed. 1997. *The ACS style guide: A manual for authors and editors,* 2nd ed. Washington, DC: American Chemical Society.

Ebel, Hans F., Claus Bliefert, and William E. Russey. 1990. *The art of scientific writing: From student reports to professional publications in chemistry and related fields.* Weinheim, Germany: VCH.

Fields, R. Douglas. 2004. The other half of the brain. *Scientific American* 290, no. 4: 54–61.

Fisher, Hans, Alycia Halladay, Nagarani Ramasubramaniam, James C. Petrucci, Dennis Dagounis, Anna Sekowski, Joseph V. Martin, and George C. Wagner. 2002. Liver fat and plasma ethanol are sharply lower in rats fed ethanol in conjunction with high carbohydrate compared with high fats. *Journal of Nutrition* 132, no. 9: 2732–2736.

Fitzpatrick, Jacqueline, Jan Secrist, and Debra J. Wright. 1998. *Secrets for a successful dissertation.* Thousand Oaks, CA: Sage.

Fleischmann, Martin, and Stanley Pons. 1989. Electromagnetically-induced nuclear fusion in deuterium. *Journal of Electroanalytical Chemistry* 261: 301–308.

Flesch, R. 1949. *The art of readable writing.* New York: Harper.

Freedman, Morris. 1958. The seven sins of technical writing. *College Composition and Communication* 9, no. 1: 10–16.

Friedland, Andrew J., and Carol L. Folt. 2000. *Writing successful science proposals.* New Haven: Yale University Press.

Friedman, Sharon M., Rae Goodell, and Lawrence Verbit. 1978. *Directory of science communication courses and programs.* Binghamton, NY: Department of Chemistry, SUNY at Binghamton.

Galazka, A. M., S. E. Robertson, and A. Kraigher. 1999. Mumps and mumps vaccine: A global view. *Bulletin of the World Health Organization* 77, no. 1: 3–14.

Garson, G. David. 2002. *Guide to writing empirical papers, theses, and dissertations.* New York: Marcel Dekker.

Gibaldi, Joseph. 2003. *MLA handbook for writers of research papers,* 6th ed. New York: The Modern Language Association of America.

Glass, Bentley. 1965. The ethical basis of science. *Science* 150, no. 3701: 1254–1261.

Goldbort, Joanne, and Dena Cochran. 2003. Development of a postpartum depression consensus statement through partnering to improve maternal-infant outcomes. Poster presented at AWHONN National Conference, Milwaukee, WI.

Goldbort, R. C. 1975. A study of the butanediols as an approach to understanding the relationship of alcohol tolerance to alcohol preference in inbred strains of mice. MS thesis, Indiana University of Pennsylvania.

Goldbort, R. 1995. Scientific information in cyberspace. *National Forum* 75, no. 2: 8–9.

Goldbort, R., and R. Hartline. 1975. Selection of butanediols by inbred mouse strains: Differences in specific activity and central nervous system sensitivity. *Federation Proceedings* 34, no. 3: 720.

Goldbort, R., C. W. Schneider, and R. Hartline. 1976. Butanediols: Selection, open field activity and NAD reduction by liver extracts in inbred mouse strains. *Pharmacology, Biochemistry, and Behavior* 5, no. 3: 263–268.

Gopen, George D., and Judith A. Swan. 1990. The science of scientific writing. *American Scientist* 78, no. 6: 550–558.

Gould, Stephen Jay. 1977. Biological potentiality vs. biological determinism. In *Ever since Darwin: Reflections in natural history.* New York: Norton, 251–259.

———. 1980. Women's brains. In *The panda's thumb: More reflections in natural history.* New York: Norton, 152–159.

Gould, S. J., and R. C. Lewontin. 1979. The spandrels of San Marco and the Panglossian paradigm: A critique of the adaptationist programme. *Proceedings of the Royal Society of London, Series B: Biological Sciences* 205, no. 1161: 581–598.

Greenly, Robert. 1993. How to write a résumé. *Technical Communication* 40, no. 1: 42–48.

Gunning, Robert. 1968. *The technique of clear writing,* rev. ed. New York: McGraw Hill.

Hall, Jeffrey C. 1994. The mating of a fly. *Science* 264, no. 5166: 1702–1714.

Halloran, S. Michael, and Annette Norris Bradford. 1984. Figures of speech in the rhetoric of science and technology. In *Essays on classical rhetoric and modern discourse,* ed. Robert J. Connors, Lisa S. Ede, and Andrea A. Lunsford. Carbondale: Southern Illinois University Press, 179–192.

Hatch, Richard. 1983. *Business writing.* Chicago: Science Research Associates.

Heath, Andrew C., and Elliot C. Nelson. 2002. Effects of the interaction between genotype and environment: Research into the genetic epidemiology of alcohol dependence. *Alcohol Research and Health* 26, no. 3: 193–201.

Hoshiko, T. 1993. Responsible conduct of scientific research: A one-semester course for graduate students. *American Journal of Physiology* 264: S8–S10.

Hunter, R. J., C. Neagoe, H. A. Jarvelainen, C. R. Martin, K. O. Lindros, W. A. Linke, and V. R. Preedy. 2003. Alcohol affects the skeletal muscle proteins, titin and nebulin in male and female rats. *The Journal of Nutrition* 133, no. 4: 1154–1157.

Isaacs, Leonard. 1986. Creation and responsibility in science: Some lessons from the modern Prometheus. In *Creativity and the imagination: Case studies from the classical age to the twentieth century,* ed. Mark E. Amsler. Newark: University of Delaware Press, 59–81.

Jansen, Johan F. G. A., Ellen M. M. de Brabander-van den Berg, and E. W. Meijer. 1994. Encapsulation of guest molecules into a dendritic box. *Science* 266, no. 5188: 1226–1229.

Jaffee, Jamison S., Phillip C. Ginsberg, Daniel M. Silverberg, and Richard C. Harkaway. 2002. The need for voiding diaries in the evaluation of men with nocturia. *Journal of the Association of Osteopathic Medicine* 102, no. 5: 261–265.

Jones, Richard Foster. 1951. Science and language in England of the mid-seventeenth century (orig. pub. 1932). In *The seventeenth century: Studies in the history of English thought and literature from Bacon to Pope.* Palo Alto: Stanford University Press.

Kanare, Howard M. 1985. *Writing the laboratory notebook.* Washington, DC: American Chemical Society.

Katz, Michael J. 1985. *Elements of the scientific paper.* New Haven: Yale University Press.

Klein, Karen Potvin, ed. 1999. *Editing grant proposals.* CBE guidelines, no. 3 (general ed. Miriam Bloom). Reston, VA: Council of Biology Editors.

Kuhn, Thomas. 1970. *The structure of scientific revolutions.* Chicago: University of Chicago Press.

Locke, David. 1992. *Science as writing.* New Haven: Yale University Press.

Lucent Technologies. 2002. Report of the investigation committee on the possibility of scientific misconduct in the work of Hendrik Schön and coauthors. Distributed by the American Physical Society, http://publish.aps.org/reports/lucentrep.pdf (retrieved October 19, 2004).

Luria, S. A. 1984. *A slot machine, a broken test tube: An autobiography.* New York: Harper and Row.

Matthews, Diane L. 1990. The scientific poster: Guidelines for effective visual communication. *Technical Communication* 37, no. 3: 225–232.

Matthews, Janice R., John M. Bowen, and Robert W. Matthews. 1996. *Successful scientific writing.* Cambridge: Cambridge University Press.

McMillan, Victoria E. 1997. *Writing papers in the biological sciences,* 2nd ed. Boston: Bedford.

Medawar, P. B. 1979. *Advice to a young scientist.* New York: Harper and Row.

Mello, M. M., and T. A. Brennan. 2003. Due process in investigations of research misconduct. *New England Journal of Medicine* 349, no. 13: 1280–1286.

Mohanty, Udayan, Irwin Oppenheim, and Clifford H. Taubes. 1994. Low-temperature relaxation and entropic barriers in supercooled liquids. *Science* 266, no. 5184: 425–427.

Montalto, Norman J. 2002. Recommendations for the treatment of nicotine dependency. *Journal of the American Osteopathic Association* 102, no. 6: 342–348.

Montgomery, Scott L. 2003. *The Chicago guide to communicating science.* Chicago: University of Chicago Press.

Morimoto, Michio, Ronald C. Reitz, Robert J. Morin, and Khoa Nguyen. 1995. CYP-2E1 inhibitors partially ameliorate the changes in hepatic fatty acid composition induced in rats by chronic administration of ethanol and a high fat diet. *Journal of Nutrition* 125, no. 12: 2953–2964.

Myers, Greg. 1990. *Writing biology: Texts in the social construction of scientific knowledge.* Science and Literature Series, general ed. George Levine. Madison: University of Wisconsin Press.

Naimi, Timothy, Robert D. Brewer, Ali Mokdad, Clark Denny, Mary K. Serdula, and James S. Marks. 2003. Binge drinking among US adults. *Journal of the American Medical Association* 289, no. 1: 70–75.

National Academy of Sciences (NAS), Committee on Science, Engineering, and Public Policy. 1995. *On being a scientist: Responsible conduct in research.* Washington, DC: National Academy Press.

National Library of Medicine recommended formats for bibliographic citation, supplement: Internet formats. 2001. Bethesda, MD: United States Department of Health and Human Services, Public Health Service, National Institutes of Health. http://www.nlm.nih.gov/pubs/format/internet.pdf (retrieved May 20, 2003).

Niven, W. D., ed. 1890. *The scientific papers of J. C. Maxwell,* vol. 2. Cambridge: Cambridge University Press.

Ogden, Thomas E., and Israel A. Goldberg. 2002. *Research proposals: A guide to success,* 3rd ed. San Diego: Academic Press.

Pascal, Chris B. 2000. Scientific misconduct and research integrity for the bench scientist. *Proceedings of the Society for Experimental Biology and Medicine* 224: 220–230.

Perelman, Leslie C., James Paradis, and Edward Barrett. 1998. *The Mayfield handbook of technical and scientific writing.* Mountain View, CA: Mayfield.

Pridemore, William A. 2002. Vodka and violence: Alcohol consumption and homicide rates in Russia. *American Journal of Public Health* 92, no. 12: 1921–1930.

Pyrczak, Fred. 2000. *Completing your thesis or dissertation: Professors share their techniques and strategies.* Los Angeles: Pyrczak's Publishing, 2000.

Ramstedt, Mats. 2002. Alcohol-related mortality in 15 European countries in the postwar period. *European Journal of Population* 18, no. 4: 307–323.

Redish, Janice C. 1985. The plain English movement. In *The English language today,* ed. S. Greenbaum. New York: Pergamon.

Schneider, Carl W., S. K. Evans, and M. B. Chenoweth. 1972. Reduction in ethanol self-selection of C57BL/6j mice during treatment with 3-((2-imidazoline-2yl) methyl) indole (36645). *Proceedings of the Society for Experimental Biology and Medicine* 140, no. 4: 1221–1223.

Schneider, Carl W., Paul Trzil, and Rita D'Andrea. 1974. Neural tolerance in high and low ethanol selecting mouse strains. *Pharmacology, Biochemistry, and Behavior* 2: 549–551.

Schowen, K. Barbara. 1997. Tips for effective poster presentations. In *The ACS style guide: A manual for authors and editors,* ed. Janet S. Dodd. Washington, DC: American Chemical Society, 27–38.

Selzer, Jack, ed. 1993. *Understanding scientific prose.* Madison: University of Wisconsin Press.

Snow, C. P. 1934. *The search.* New York: New American Library.

Spedding, James, Robert Leslie Ellis, and Douglas Denon Heath, eds. 1876–1883. *The bookes of Francis Bacon, of the proficience and advancement of learning, twoo divine and humane* (orig. pub. 1605). In *The works of Francis Bacon,* vol. 6. Boston: Houghton, Mifflin.

Sprat, Thomas. 1958. *History of the Royal Society* (orig. pub. 1667), ed. J. I. Cope and H. W. Jones. St. Louis: Washington University Press.

Stanley, S. M., and X. Yang. 1994. A double mass extinction at the end of the Paleozoic era. *Science* 266, no. 5189: 1340–1344.

Strange, A. W, C. W. Schneider, and R. Goldbort. 1976. Selection of C_3 alcohols by high and low ethanol selecting mouse strains and the effects on open field activity. *Pharmacology, Biochemistry, and Behavior* 4, no. 5: 527–530.

Tampier, L., and M. E. Quintanilla. 2002. Effect of acetaldehyde on acute tolerance and ethanol consumption in drinker and nondrinker rats. *Journal of Studies on Alcohol* 63, no. 3: 257–262.

Thomas, Lewis. 1979. Notes on punctuation. In *The medusa and the snail: More notes of a biology watcher.* New York: Viking, 125–129.

Toulmin, S. E., R. Rieke, and A. Janik. 1978. *An introduction to reasoning.* New York: Macmillan.

Tufte, E. R. 1997. *Visual explanations.* Cheshire, CT: Graphics Press.

Verbit, Lawrence P. 1983. *Directory of science communication courses, programs, and faculty,* 2nd ed. New York: SUNY-Binghamton, Department of Chemistry.

Walker, Lorraine O. 1996. Predictors of weight gain at 6 and 18 months after child-birth: A pilot study. *Journal of Obstetric, Gynecologic, and Neonatal Nursing* 25, no. 1: 39–48.

Watson, James D. 1968. *The double helix.* New York: New American Library.

Watson, J. D., and H. F. C. Crick. 1953. A structure for deoxyribose nucleic acid. *Nature* 346, no. 4356: 737–738.

Watson, J. D., and H. F. C. Crick. 1980. The structure of DNA. In *The double helix,* ed. Gunther S. Stent. New York: Norton, 257–274 (orig. pub. in *Cold Spring Harbor Symposia on Quantitative Biology* 18, no. 1953: 123–131).

Weaver, David, Moema H. Reis, Christopher Albanese, Frank Costantini, David Baltimore, and Thereza Imanishi-Kari. 1986. Altered repertoire of endogenous immunoglobulin gene expression in transgenic mice containing a rearranged Mu heavy chain gene. *Cell* 45: 247–259. (Retracted in 1991, *Cell* 59: 536.)

Wells, James P., David L. Hyler-Both, Tabitha D. Danley, and Gregory H. Wallace. 2002. Biomechanics of growth and development in the healthy human infant: A pilot study. *Journal of the American Osteopathic Association* 102, no. 6: 313–319.

Whitburn, Merrill D. 1978. The plain style in scientific and technical writing. *Journal of Technical Writing and Communication* 8, no. 4: 349–358.

Wilkinson, Antoinette M. 1991. *The scientist's handbook for writing papers and dissertations.* Englewood Cliffs, NJ: Prentice Hall.

Willey, Basil. 1953. *The seventeenth-century background* (orig. pub. 1934). New York: Doubleday.

Wilson, Edward O. 1975. *Sociobiology: The new synthesis.* Cambridge: Harvard University Press.

Wolff, R. S., and L. Yeager. 1993. *Visualization of natural phenomena.* New York: Springer-Verlag.

Woodford, F. Peter, ed. 1986. *Scientific writing for graduate students.* Bethesda, MD: Committee on Graduate Training in Scientific Writing, Council of Biology Editors.

———. 1999. *How to teach scientific communication.* Reston, VA: Council of Biology Editors.

Yamamoto, Takashi. 2003. Brain mechanisms of sweetness and palatability of sugars. *Nutrition Reviews,* 61, no. 5: S5–S9.

INDEX